Some Facts About Ore Deposits

by Arizona Bureau of Mines

with an introduction by Kerby Jackson

Introduction

It has been decades since the Arizona Bureau of Mines released their important publication "Some Facts About Ore Deposits". First released in 1935, this important volume has been out of print and has been unavailable to the mining community since those days, with the exception of expensive original collector's copies and poorly produced digital editions.

It has often been said that *"gold is where you find it"*, but even beginning prospectors understand that their chances for finding something of value in the earth or in the streams of the Golden West are dramatically increased by going back to those places where gold and other minerals were once mined by our forerunners. Despite this, much of the contemporary information on local mining history that is currently available is mostly a result of mere local folklore and persistent rumors of major strikes, the details and facts of which, have long been distorted. Long gone are the old timers and with them, the days of first hand knowledge of the mines of the area and how they operated. Also long gone are most of their notes, their assay reports, their mine maps and personal scrapbooks, along with most of the surveys and reports that were performed for them by private and government geologists. Even published books such as this one are often retired to the local landfill or backyard burn pile by the descendents of those old timers and disappear at an alarming rate. Despite the fact that we live in the so-called "Information Age" where information is supposedly only the push of a button on a keyboard away, true insight into mining properties remains illusive and hard to come by, even to those of us who seek out this sort of information as if our lives depend upon it. Without this type of information readily available to the average independent miner, there is little hope that our metal mining industry will ever recover.

This important volume and others like it, are being presented in their entirety again, in the hope that the average prospector will no longer stumble through the overgrown hills and the tailing strewn creeks without being well informed enough to have a chance to succeed at his ventures.

Kerby Jackson
Josephine County, Oregon
May 2016

PREFACE

This little book consists principally of a series of articles that appeared during 1919 and 1920 in the *Arizona Mining Journal* (now the *Mining Journal*), published at Phoenix, Arizona.* Each chapter represents one article, and, in most instances, no material changes have been made in these articles, although the last two chapters and the appendices are new and some material has been added to each of the other chapters in order to bring them up to date.

The series was originally written for prospectors or miners who are interested in the theoretical aspects of their vocation, and who are honestly seeking to learn the truth. So many mining engineers, geologists, and students of these subjects have, however, commented favorably upon the articles and expressed a desire to obtain them in book form that it is hoped they will prove to have some value to a wider audience than that for which they were originally written.

The statements made are not original contributions to knowledge, and rarely represent merely the personal opinions and ideas of the writer. In most instances the convictions expressed are the common heritage of all students of geology and mining, and may be accepted unquestionably as representing the concensus of opinion of the great majority of such men. The assertions set forth cannot all be proven, and no attempt to prove them is made; they are the fruit of diligent and careful observations made by thousands of scientists and mining engineers scattered all over the world. While time may prove some of them to be incorrect, and will doubtless add much to our knowledge of several of the points discussed, most of the views presented may be accepted as having become so firmly established that there is small possibility that they will ever be successfully assailed. It has been necessary to indulge in broad generalizations, and it is admitted that numerous exceptions exist to specific rules given. It is believed, however, that it would prove confusing to discuss such exceptions.

Most of the chapters were originally published as part of a series entitled, "Some Common Mining Fallacies." It has been the aim of the writer, however, not so much to lay emphasis upon false ideas as to outline the truth as concisely and simply as possible. Progress cannot be made by destroying, unless something better than the thing destroyed is created, and this statement applies as truly to beliefs as to material things. The topic

*Reprinted by permission of Chas. F. Willis, Publisher.

"Mining Fallacies" formed, then, merely a pretext for the presentation of facts, or what the majority of mining geologists believe to be facts; and no apology seems necessary when comparatively little is said about fallacies and much about facts, as is the case in several of the chapters.

It is realized that some of the topics very briefly presented herewith might profitably be greatly expanded, but it was not the desire of the writer to trespass upon the field of several excellent and exhaustive treatises already in print, to which reference is made in the bibliography. It is hoped that many readers will be led to seek in these works further light on subjects of such interest and importance to our mineral industries.

While, as previously stated, almost every fact herein presented has already appeared in print, most of the chapters were written from memory without immediate reference to the literature of the subject, and it would require a vast amount of time and trouble to locate the sources of the data set forth. The writer sincerely regrets that he is not in a position to give proper credit for his ideas, and offers his apologies to anyone who may feel that his work has been slighted.

Doctor B. S. Butler and Doctor R. J. Leonard have read the manuscript critically, and the writer has adopted a number of suggestions made by them. His sincere thanks are extended to them for their invaluable assistance and encouragement.

G. M. BUTLER.

August 15, 1935.

TABLE OF CONTENTS

SOME FACTS ABOUT ORE DEPOSITS

G. Montague Butler

CHAPTER I

THE HIGHER VALUES WITH INCREASED DEPTH MYTH

A Widespread, but Fallacious, Belief

Most prospectors have an idea, so firmly established that it amounts to a conviction, that the grade of ore is almost certain to improve with increase in depth. Over and over again one hears statements to the effect that "I know the ore is low-grade here, but, if I sink a few score or hundred feet, I shall certainly find rich ore." Millions of dollars, the active portions of many men's lives, and tremendous amounts of energy have been wasted by sinking on ore deposits that a trained mining geologist would recognize as unpromising in the belief that action would certainly result in the discovery of good ore; and the hills are dotted with dumps and idle mills that are nothing but monuments to this fallacious doctrine. For, in its general application, it is undeniably fallacious and cannot be justified by either theoretical reasoning or experience.

Surely, it may be said, a belief so widely held as the one under consideration must have some basis in fact—must contain some grains of truth. Yes, the truth of this statement must be admitted, and it is the aim of this chapter to sift the wheat from the chaff—to indicate under what conditions ore may and when it may not be expected to become richer with greater depth.

The Deposition of Ore

Every prospector and miner should know that nearly all deposits of ore (and gangue minerals, as well) were formed by deposition from water solutions, usually hot or warm, and occasionally so hot as to be in the form of a gas, but sometimes cold, occasionally with a very high proportion of mineral matter in solution and, in other instances, in no way different from spring waters which contain various proportions of many substances including even metallic salts.

Such solutions may have come from various sources. In a few instances, rainwater may have sunk into the earth and penetrated horizontally, or roughly so, through mineralized

7

rocks from which it obtained its mineral content, to the points where deposition occurred; some solutions, again originally rainwater, have come from above, as usually occurs when secondary enrichments (see Chapter IV) are formed; but, in most instances, the solutions have doubtless arisen from profound depths.

There are many evidences of the validity of the theory that large quantities of hot solutions are expelled from molten earth materials (called magmas) while they are cooling and solidifying into igneous rocks, and there is good reason to believe that a considerable proportion of the many valuable metals originally contained in such molten material is excluded or expelled from it as it solidifies and that these valuable metals are carried out therefrom in hot solutions. These solutions, although composed very largely of water, also contain a number of other substances that add materially to their solvent power and enable them to hold in solution the elements or compounds later deposited as ore or gangue minerals. Furthermore, it should be remembered that the solvent power of these solutions is much increased at great depth by the high temperatures and pressures that prevail there.

It has already been stated that the elements or compounds now found in ore deposits usually once formed a part of and escaped from masses of molten earth material in solutions expelled from these cooling masses, but in some instances it is possible that the solutions became further charged with mineral matter by dissolving certain substances present in solid rocks with which the rising solutions came in contact.

Solutions that originate in the manner outlined naturally find their way upward through fissures, faults, sheared or crushed zones, or along contacts or other lines of weakness; and from these solutions the various ore and gangue minerals are deposited as a result of cooling, relief of pressure, influence of peculiar wall rocks, mingling of chemically different solutions, and other factors. Of all the conditions that bring about mineral deposition, lowering of the temperature of and decrease of the pressure on the solutions are doubtless most generally effective.

Most geologists believe that a great majority of all primary ore deposits originated as has just been outlined, and that, if it were possible to trace most veins indefinitely downward, they would be found to terminate in great masses of solidified magma —igneous rocks. The fact that most ore deposits are situated in areas where igneous rocks are plentiful, rather than in districts where only unintruded sediments are found, indicates that this hypothesis is a reasonable one.

It is true that, as a result of the factors just mentioned, deposition of some elements may begin far below the earth's surface and may continue through a vertical distance of thousands of feet. This has occurred in the free-gold-quartz or the gold-pyrite-quartz lodes. Some profitable although low-grade ore bodies of this kind have been found, which it is believed were originally deposited at a depth of as much as 10,000 feet or even more. In the case of some other minerals, such as zinc sulphide (sphalerite or blende), the vertical range of deposition, while relatively great, is less than in the instances just mentioned; and this range is still less where compounds of sulphur, arsenic, antimony, or tellurium with silver, mercury, lead, or gold are involved. In all the instances already mentioned, the deposits were doubtless originally richest near the surface, and contained ore of lower and lower grade at increasing depths where the agents causing precipitation were less effective. Most mercury, lead, and silver deposits show a notable diminution in grade in a few hundred feet, but, in other places, this decrease, while still appreciable, is less marked.

Most sulphide ores of copper and molybdenum and nearly all of the important tungsten and tin minerals were, however, probably deposited in greatest quantities at considerable depths below the surface where the temperatures were relatively high. Both above and below these points, a more or less marked diminution in grade is to be expected, but it should not be forgotten that, even under such conditions, erosion (the disintegration and removal of outcrops by wind and water) has doubtless, in a majority of instances, carried away the lower grade ores originally formed above the points of maximum deposition.

What has been said refers to the so-called "primary" ores; that is, those originally deposited at points where no ore previously existed. Such deposits do not, excepting under unusual circumstances, owe their existence to the action of downward percolating surface waters that attack some minerals and form powerful solvents that dissolve some metals and then precipitate their metallic contents elsewhere, often at lower depths, where these metals are deposited as so-called "secondary" deposits. The primary minerals are predominantly sulphides, while the secondary ores contain variable amounts of oxides, carbonates, silicates, etc., along with large or small proportions of sulphides.

Ore Deposits Usually Decrease in Value with Increasing Depth

From what has been set forth, it should be evident that, excepting under peculiar conditions yet to be discussed, most primary ore deposits show decreased values with increased depth, and that this decrease is pronounced through vertical distances of a few hundred feet in some places, while in other localities it is very gradual through depths aggregating several thousand feet.

The statements just made, while in accord with physical and chemical theory, should not be branded "theoretical," a term that is, unfortunately, regarded with disapproval by many prospectors; the conclusions reached have been corroborated during mining operations in such a great number of places that there is no doubt of the very wide applicability of the statement that a primary ore deposit should be expected to yield poorer instead of richer ore as the depth of the workings increases.

Explanation of the Prevalence of the Fallacy

The widely held belief that sinking on a deposit is almost sure to expose better ore than that found at or near the surface is probably due to three causes, namely: (1) the not uncommon existence of secondary enrichments below, or not far from, the surface; (2) the occasional presence at moderate depths of ore shoots that do not outcrop; and (3) the fact that a man is very willing to be convinced of the truth of anything long and ardently desired. While further and more extensive reference to the first two points will be made in other articles of this series, a brief discussion of each follows:

(1) It has already been mentioned that surface water will attack some minerals and form powerful solvents that dissolve some or all of the metals in many primary ore minerals. The solutions so formed often work downward through the resulting more or less leached material until conditions are encountered that cause these solutions to deposit secondary ore minerals. In regions of relatively heavy rainfall, this deposition ordinarily occurs at or near the ground-water level—that depth below which for indefinite, but often great, distances the earth is permanently saturated with water. Ores of copper are especially apt to be thus dissolved and redeposited, but secondary silver and zinc deposits are not uncommon, and other metals more rarely undergo concentration in this way.

Secondary deposits of this kind, beneath leached material, are occasionally very rich—of much higher grade than the primary ores from which they were derived; but the depth to which the enriched zone extends in each case is small, usually well under 100 feet except in "porphyry" copper deposits where it may reach 200 feet or more. The outcrop features indicating the probable presence of enriched deposits will be described in a subsequent chapter.

Now, for untold ages, prospectors have sunk through leached outcrops, and, when their work has been carried to sufficient depths, have occasionally reached such enriched zones. There, their activities have usually been suspended for two reasons: first, prospectors are rarely satisfied to settle down to the relatively unexciting work of mining ore which they have exposed; and, second, the cost of handling the water that saturates most secondary deposits is beyond the means of the majority of prospectors. They, therefore, under such conditions, usually dispose of their properties without attempting to sink further, and move on to fresh fields. This has happened so frequently that the idea that all deposits become richer at depth has, not unnaturally, become firmly established. When no rich ore is found it is usually supposed that the depth attained has been insufficient to expose it. Since the prospector is rarely able to sink through the enriched zone, he is unaware of the sudden decrease in values that is bound to occur at a limited distance below the top of the secondarily enriched deposit, and does not know that the primary ore below, even if of sufficiently high grade to yield a profit, will become leaner with more or less rapidity as greater depth is attained.

(2)   Practically all prospectors are familiar with the fact that in most types of ore deposits the primary ore is not of uniform grade at all points. They know that "shoots" of rich ore are apt to be scattered through the leaner material, and that the distance between these shoots, and their size and shape, vary greatly.

The position of some ore shoots is evidently determined by the proximity of certain kinds of rocks, the presence of an intersecting fissure, the existence of an unusually shattered and porous condition of the rock through which the solutions ascended, or other factors; but in most deposits the distribution of such shoots, especially in veins, cannot be explained.

Now, if a primary ore shoot of this kind fails by a few score or hundreds of feet to reach the surface, a miner, sinking through the lean ore above, would note a sudden or gradual increase in the grade and, probably, in the quantity of his output. This has doubtless happened in a sufficient number of instances to give color to the fallacy under consideration. One or two cases of this kind would be long remembered and information concerning them widely disseminated, while hundreds of failures would attract little attention, since negative results could easily be attributed to failure to sink far enough.

Since the chance of striking such a non-outcropping shoot in most deposits is extremely small, it would be folly to sink in lean surface ore with this possibility in mind.

(3)   The last explanation suggested, that hope is the father of belief, may seem at first thought rather unsatisfactory, but there is no doubt of its validity. What men hope to be true and what they desire to believe is very apt to become a conviction, as witness the almost universal expectation of life beyond the grave. Prospectors hope and desire that lean ore on the surface will become richer as they sink, and this idea seems to be justified by the fact that, in some places, secondary enrichments or non-outcropping ore shoots have been encountered; so it is not surprising that the erroneous belief that sinking will almost surely expose richer ore has become firmly established. It is an undoubted fact, though, that the chances are decidedly in favor of primary ore becoming leaner rather than richer as greater depths are attained.

Conclusion

A certain prospectus contained the statement that "it is believed that the property will become richer and the ore quantities greater as additional depth is gained." The editor of the *Mining and Scientific Press*, in commenting on this phrase, said in the June 14, 1919, issue: "That is an old nursery tale, a pretty fantasy, and on a par with the tale of the treasure at the rainbow's end."

From scores of similar quotations bearing on the matter under consideration that might be given, the following have been selected as especially pertinent.

Waldemar Lindgren, in *Mineral Deposits*, writes: "If this [decreasing pressure stands first among the conditions favoring precipitation] is true, the deposits should generally become poorer or barren in depth. In a general way this is doubtless true." The same writer, in *Economic Geology*, Volume I, page 34, states: "As men begin to delve deeper into the earth's crust instead of confining themselves to the shallow diggings and mines of earlier times, many deposits are found to diminish in size or value."

J. P. Wallace in *A Study of Ore Deposits for the Practical Miner*, says: "The notion so often preached that one has only to sink to get good ore is a delusion; it is against the experience of good miners the world over, and is rarely verified," and, "should no outcrop of good ore be found, it is best for the prospector with little means to allow some one else to do the sinking."

Finally, it should be remembered that it costs more to mine ore from deep than from shallow deposits, so, even though the value of the ore does not diminish notably as depth is attained, operations may become increasingly unprofitable with deeper mining and finally cost more than the product is worth.

CHAPTER II

THE CHANGE OF CHARACTER OF ORE WITH INCREASED DEPTH MYTH

In the preceding chapter an attempt was made to show that, excepting under certain peculiar conditions, there is no justification for the hope that relatively low-grade ore at or near the earth's surface will give place at greater depth to ore that is more valuable simply because it contains a larger proportion of the same metals as are found above. Although not so generally held, many miners adhere to the idea that ore of little or no value found near the surface will prove transitional at greater depth into material that is valuable because it contains metals different from those found in the shallower workings. For instance, the belief may be expressed that a zinc-lead ore will change gradually or suddenly into one containing chiefly lead and silver; or a quartz lode that contains primary iron sulphide and carries a trace of copper will show a largely increased copper content with deeper mining. Claims of this kind are frequently encountered in the prospectuses of companies organized to work relatively undeveloped prospects, and, while such claims are probably in some cases known to be unjustifiable and are advanced with the intent to deceive prospective purchasers of stock, in many instances they are merely the manifestation of undue optimism and ignorance. In either case development of such properties is quite likely to lead to disappointment and to cause losses that serve to magnify unnecessarily the risks involved in mining operations.

The Facts

It is undeniable that the nature of the ore in a given lode may change as greater depth is attained. An iron oxide outcrop may overlie a rich deposit of copper-iron sulphide; a nearly barren surface exposure may give place at relatively shallow depth to a high grade deposit of zinc carbonate underlain by zinc sulphide; an outcrop stained with oxides of iron and manganese may, under certain conditions, lead to a valuable deposit of gold when followed downward. Such conditions are, however, always the result of leaching by surface waters, and subsequent deposition at lower levels of metals carried downward in solution. Such valuable ore deposits constitute the secondary enrichments mentioned briefly in the preceding chapter and to be discussed at some length later.

Does the character of *primary ore* ever change with increased depth? Most assuredly, but a zinc-lead ore will not prove transitional into one containing chiefly lead and silver at greater depths, and unleached primary iron sulphide will not show an increase of copper content when followed downward.

Are not, then, such changes more or less matters of accident or chance and quite unpredictable? By no means; they follow certain well established and relatively simple general laws that are not commonly understood by prospectors and will now be outlined.

The Primary Ore Zones

In 1907 Waldemar Lindgren[1] and J. E. Spurr[2] were the first persons in the United States to advance the theory that the character of the metals deposited from rising solutions is dependent upon the depth and probably even more directly upon the temperature at which such deposition takes place. As a result of the work of these two geologists and other investigators, it may now be stated, without fear of contradiction, that, if several different metals are present in a solution rising through the earth, the deposition of all will not begin at the same point, but, on the contrary, ores of a certain metal will be precipitated at the greatest depth, ores of another metal may be formed above this point, and ores of a third metal may possibly be deposited still higher. In the light of our present knowledge it is possible to divide deposits of primary ore minerals according to the depth or temperature at which they were formed into a number of more or less distinct zones. The most important and widely recognized of such zones, according to Spurr,[3] with the uppermost at the top of the list, are as follows:

The silver zone
The lead (galena) zone
The zinc (low-iron sphalerite or blende) zone
The arsenic (arsenopyrite) zone
The copper (chalcopyrite) zone
The iron (pyrite) and gold zone
The rare metals or pegmatite zone

In order to grasp the idea involved in the preceding statements, assume that cassiterite (tin oxide), pyrite, chalcopyrite, arsenopyrite, low-iron sphalerite, galena, and argentite (silver glance or silver sulphide) are all present in a rising solution. Tin oxide should be deposited at the greatest depth, most of the pyrite higher, the largest portion of the chalcopyrite still higher, and so on through the list in the order in which the minerals are named.

It matters not that the illustration given is purely hypothetical, that it is extremely doubtful whether any solution would con-

---

[1] *Economic Geology*, Vol. II, p. 105.
[2] *Ibid.*, p. 781.
[3] "Theory of Ore Deposition," J. E. Spurr, *Economic Geology*, August, 1912, pp. 485-492.

tain all the minerals mentioned, and that the example is probably representative of no existing ore deposit, it is the relative vertical position that minerals of two or more metals normally occupy with reference to each other that is important, and it is, generally speaking, as shown in the list of ore zones.

Since this list shows the vertical order in which the metals occur from the surface downward, it should be evident that one of the lower should not generally overlie one of the upper zones; that, if valuable deposits of sphalerite (which contains little or no iron) and chalcopyrite are present in the same vein, the copper will probably underlie the zinc.

Some knowledge of the theory under consideration will enable one to predict changes in the character of primary ore with increased depth in a manner that appears miraculous to the uninitiated; but in order to make such forecasts it will be necessary to possess further information concerning the character of each zone, to consider certain exceptions to the general rules laid down, and to discuss and illustrate the practical application of these rules.

The Silver Zone.—The ore present in this zone usually consists largely of easily dissolved or volatilized compounds. Although silver is commonly the most abundant valuable metal represented in this zone, it may contain much gold (usually alloyed with the silver), as well as varying amounts of antimony, arsenic, bismuth, tellurium, and selenium. Mercury ores which rarely contain silver, while rare, are largely, although not completely, confined to this zone. The free gold in this zone is apt to be alloyed with so much silver as to be pale yellow in color and it frequently occurs in fine particles.

The common ore minerals are argentite (silver sulphide), silver-bearing gray copper (copper sulpharsenite or sulphantimonite), a number of arsenical and antimonial compounds of silver, and stibnite (antimony sulphide). Many other minerals mentioned later as characteristic of deeper zones may also be present in subordinate amounts; and a rather complex mixture of several different ore minerals is a fairly common feature of this zone.

Quartz is probably the commonest gangue mineral; in fact, the deposits in this zone characteristically take the form of quartz veins. Other gangue minerals of lesser importance are opal (hydrous silica), chalcedony (silica), calcite (lime carbonate), adularia (aluminum-potassium silicate), and, rarely or locally, fluorite (calcium fluoride), barite (barium sulphate), and rhodochrosite (manganese carbonate). The gangue is, characteristically, far in excess of the ore minerals.

This zone is a very important source of gold and silver and comprises many of the great unenriched bonanzas of these metals such as are commonly associated with flows or intrusions of comparatively recent igneous rocks.

The Lead Zone.—Galena (lead sulphide), which may or may not contain silver, is the most plentiful mineral in this zone.

Occasionally the veins or other types of deposits may contain little or no gangue, but in other deposits, a variety of minerals, such as barite (barium sulphate), fluorite (calcium fluoride), rhodochrosite (manganese carbonate), calcite (calcium carbonate), ankerite (calcium-iron-mangnesium carbonate), siderite (iron carbonate), and quartz, may make up a plentiful gangue. This is probably the principal manganese zone.

Zinc Zone.—Excepting that zinc sulphide (sphalerite or blende), comparatively free from iron, is the dominant mineral and the gangue is more apt to be largely quartz, in this zone, it is almost identical with the lead zone, and in fact they are frequently commingled. In such cases, however, the lead alone may be found above the point where lead and zinc are commingled and the zinc is almost sure to persist to greater depths than the lead. When the proportion of sphalerite is much greater than that of galena, silver is not usually very plentiful. The sphalerite may or may not contain gold.

Arsenic Zone.—Arsenic is the characteristic metal or semi-metal of this zone, but nickel, cobalt, and silver (sometimes in the form of "native" silver) are plentiful in some localities, and antimony may also occur in the ore.

Arsenopyrite, which may carry gold, is the commest ore mineral and may or may not contain silver, but in some localities it is accompanied by, or largely replaced with, pyrite (iron sulphide) which may carry silver. Pyrrhotite (iron monosulphide) is also sometimes present. The nickel and cobalt are usually represented by cobaltite (cobalt arsenide and sulphide), smaltite (cobalt-nickel arsenide), or intermediate or related species.

The ore may completely fill the veins or other types of deposit, or a plentiful gangue (usually largely quartz) may be present.

This zone occasionally appears to underlie the copper zone and is not so commonly developed as are most of the others mentioned, but deposits belonging to it are occasionally of great economic importance, as at Cobalt, Ontario, Canada. When it apparently underlies the copper zone it may contain tungsten.

Copper Zone.—Copper is the principal metal found in this zone, but it may be accompanied by subordinate amounts of gold and silver. High-iron sphalerite (zinc sulphide) may also occur in this zone.

Chalcopyrite is the common ore mineral, but primary chalcocite (copper sulphide), bornite (copper-iron sulphide), and other primary copper minerals and pyrrhotite (iron monosulphide) are sometimes present in appreciable quantities.

The gangue is usually largely quartz, but carbonates are sometimes present, and, in a number of important deposits, the ore occurs as an impregnation or dissemination in porphyry or some other type of igneous rock or as a replacement of schist.

Although this zone is the principal source of copper, the ore is usually of relatively low grade unless it has been subjected to secondary enrichment.

The Iron-Gold Zone.—This zone contains the great, compact quartz veins that carry more or less pyrite which often contains some gold, free gold, or both. It is the principal gold (containing little silver) zone, although the greatest bonanzas of this metal are apt to be in the uppermost zone.

The Rare Metals or Pegmatite Zone.—While this zone is often barren it may contain tin, molybdenum, tungsten, bismuth, gems, monazite, zircon, etc.

In true pegmatites the gangue usually consists predominantly of quartz, orthoclase, and micas with or without varying proportions of tourmaline, topaz, beryl, titanite, apatite, etc.; and the gangue usually greatly predominates over the ore or other valuable mineral present in all types of deposits in this zone. In non-pegmatic deposits in the pegmatite zone, the gangue is usually mainly quartz.

This zone is distinguished by the rare metals and minerals that it contains, although some deposits of these substances do not belong to it. It furnishes wonderful mineral specimens for museums, and yields in some places commercially important amounts of the various metals and minerals mentioned, but, in the great majority of instances, the pegmatite dikes or veins are decidedly barren.

Other Facts Relating to the Primary Ore Zones

Some authorities would add two additional zones to those already given. One, the surface zone, is rarely of economic importance. It includes the hot-spring deposits that usually consist largely of calcite or opal silica. Fluorite, barite, celestite (strontium sulphate), and other gangue minerals may also develop at or near the surface. While ore minerals are comparatively rare in this zone, they include stibnite (antimony sulphide), marcasite (iron sulphide of the same composition as pyrite, but differing therefrom in crystallization and other physical properties), and cinnabar (mercury sulphide). The second additional ore zone mentioned may be called the magmatic zone, and includes those deposits formed by the segregation of metallic minerals originally dissolved in magmas. As the character of such minerals varies with the nature of the igneous rock with which they are associated, rather than with the depth at which they formed, further discussion of this zone seems unnecessary except to say that the most characteristic ores formed in this way are magnetite (magnetic iron oxide) and chromite (iron-chromium oxide).

It should be plainly understood that all the zones mentioned are probably never developed in any one vein or other type of deposit. Two, or possibly three, are all that have ever been exposed in any one mine. This condition exists not only because, in most instances, some of the upper zones have probably been removed by erosion, but it is also due to the fact that some of the zones are so thick that mining operations cannot be carried deep enough to expose all of the changes in character of ore that might otherwise be discovered. Further, it is an undoubted fact that several may be completely absent from some ore deposits. For instance, the zinc zone is frequently transitional downward into the copper zone without any indication of the presence of the arsenic zone. Similarly, the silver zone may be transitional downward into the iron zone in which case four possible zones are missing. If, however, a deposit that contains minerals that evidently belong to the silver zone also contains a considerable proportion of galena, it is practically certain that the proportion of galena will increase at greater depth, and that the many complex minerals characteristic of the silver zone will become less plentiful. Further, if the deposit contains a large proportion of iron-free sphalerite with considerable quantities of chalcopyrite, but little or no arsenical minerals, it may be expected that the arsenic zone is absent and that the zinc will become less plentiful and the chalcopyrite more plentiful at greater depth.

It should also be borne in mind that the zones are not sharply separated from one another. In fact the zones are gradually transitional into each other, and there is frequently a considerable vertical distance through which a deposit partakes of the characteristics of two or more different zones. As previously stated the lead and zinc zones are especially apt to be commingled in this way.

Gold and pyrite may occur in any zone to some extent, but they are very rare in the rare metals or pegmatite zone except near its top. The same statement is true of chalcopyrite, although it is almost never present in commercially important amounts in any but the copper zone.

It is, unfortunately, impossible to give any definite figures relative to the thickness of the different zones. Indeed, it is very likely that the copper zone, for instance, may be thick in one locality where the rate of decrease of temperature toward the surface was relatively low, and thin where the rate was high. It may be said, however, that available data indicate that the silver, lead, zinc, and arsenic zones are usually each measurable in hundreds of feet, while the copper, iron-gold and the rare metals or pegmatite zone are thousands of feet thick.

Emmons' Table of Ore Zones

In 1924 W. H. Emmons published a table[4] of the various ore zones that he believed could be identified, which is reproduced on the following page.

---

[4] "Primary Downward Changes in Ore Deposits," *Trans. A.I.M.E.*, 1924, pp. 964-997.

Surface

Barren	1.	Barren zone, chalcedony, quartz, barite, fluorite, etc. Some veins carry a little mercury, antimony, or arsenic.
Mercury	2.	Quicksilver veins, commonly with chalcedony, marcasite, etc. Barite-fluorite veins.
Antimony	3.	Antimony ores—stibnite often passing downward into lead, with antimonates. Many carry gold.
Gold	4.	Bonanza ores of precious metals. Argentite, antimony and arsenic minerals common.
Silver		Silver minerals, some copper, lead, and zinc sulfides, quartz, calcite, rhodochrosite, adularia, alunite, etc.
Barren	5.	Most nearly consistent barren zone, represents the bottoms of many Tertiary precious metal veins. Quartz, carbonates, etc., with pyrite and small amounts of other sulfides.
Silver	6.	Argentite veins, complex antimony-silver sulfides, stibnite, etc. Galena veins with silver. Commonly silver decreases with depth. Quartz gangue, siderite common, often increasing with depth.
Lead	7.	Galena veins, commonly with some silver. Sphalerite generally present, increasing with depth. Chalcopyrite common. Gangue is quartz and often carbonates (Fe, Mn, Ca).
Zinc	8.	Sphalerite veins with some lead, and chalcopyrite, quartz, gangue.
Copper	9.	Tetrahedrite veins, commonly argentiferous, chalcopyrite present. Some pass downward into chalcopyrite. Enargite veins generally with tetrahedrite and tennantite.
Copper	10.	Chalcopyrite veins, generally with pyrite, often with pyrrhotite. The gangue is quartz and in some places carbonates. Some pass downward into pyrite and pyrrhotite with a little chalcopyrite. Generally carry silver or gold.
Gold	11.	Gold veins with quartz, pyrite, and commonly arsenopyrite and chalcopyrite. At places zones 10 and 11 are reversed.
Bismuth	12.	Bismuthinite and native bismuth with quartz and pyrite, etc.
Arsenic	13.	Arsenopyrite with chalcopyrite and often tungsten ores.
Tungsten	14.	Tungsten veins with quartz, pyrite, chalcopyrite, pyrrhotite, etc. Arsenopyrite is commonly present.
Tin	15.	Cassiterite veins with quartz, tourmaline, topaz, etc.
Barren	16.	Quartz with small amounts of other minerals.

At first sight, this list of zones may seem to differ materially from that already given, but careful scrutiny will show that the two schemes are essentially alike although Emmons has subdivided many of Spurr's zones. The general equivalency of the zones outlined in this book and recognized by Emmons is indicated in the following table:

Zones outlined herein	Emmons' zones
Surface zone	1. Barren zone.
Silver zone	2. Mercury zone.
	3. Antimony zone.
	4. Gold and silver zone.
	5. Barren zone.
	6. Silver zone.
Lead zone	7. Lead zone.
Zinc zone	8. Zinc zone.
Arsenic zone / Copper zone	9. and 10, Copper zones; part of 11, Gold zone; and all of 13, Arsenic zone.
Iron and gold zone	11. Gold zone.
Rare metals or pegmatite zone	12. Bismuth zone.
	14. Tungsten zone.
	15. Tin zone.
	16. Barren zone.

The only serious difference is in the position of the arsenic zone which Emmons places below his copper and gold zones, instead of above the copper zone where Spurr, whose ideas have been followed in this chapter, believes it should go.

Since it is a well known fact that arsenopyrite (iron arsenide and sulphide) is often associated with tungsten ores, it is likely that there are two or more distinct arsenic zones and it may some day be possible to describe with certainty the characteristics of each of them. Until these distinctive features of the different arsenic zones have been set forth, it is probably wise to think of the arsenic zone as in some places overlying and elsewhere underlying the copper zone, and to accept Spurr's conclusion that the principal arsenic zone usually overlies the copper zone.

Emmons' barren zone (5) is particularly interesting since he points out that few of the veins in his overlying zones have been found to pass through this barren zone. He also points out that at many places his zones 7 and 8 bottom in barren quartz in granite, and that "essentially barren material—quartz with a little pyrite and other minerals—may be found below zones 6 and 14."

Arrangement of Deposits in Zones on the Surface

The ore zones under consideration are not necessarily or, in all probability, even usually roughly parallel to the surface of the earth. In any restricted area, as in one mine or a group of contiguous mines, any ore zone will doubtless roughly parallel the surface, but any ore zone in an extensive mineralized area is more apt to have the shape of a rough dome—it will approach the surface more closely somewhere toward the center of the area than it does around its edges. This condition results from the fact that the solutions from which the ore and gangue precipitate usually originate in a mass of cooling magma, deep below the surface. Such a mass of cooling magma usually has an upper surface that is roughly dome-shaped; at one point it approaches the surface much closer than elsewhere.

Now since each ore zone is apt to be deposited approximately the same distance from the source of the mineralizing solutions, it follows that a given ore zone may deposit 5,000 feet, for example, from the surface above the point where the source of the solutions comes closest to the surface, and the same zone may be 7,000 or 8,000 feet from the surface around the edges of the mineralized area where the source of the solutions is 2,000 or 3,000 feet farther from the surface. In other words the different ore zones form a series of shells which are only approximately parallel to the roughly dome-shaped upper surface of the deeply buried mass of igneous rock from which the mineralizing solutions were expelled as the magma cooled.

If such a series of roughly dome-shaped shells are exposed by erosion, a relatively deep ore zone will be exposed near the center of the area. This zone may be surrounded by a ring of outcrops of deposits characteristic of a shallower zone, and other rings of outcrops representative of still shallower zones may occur still farther out.

The ore deposits in the neighborhood of Butte, Montana, are a good illustration of the condition just described. In the central area, where the great copper mines are situated, the copper zone comes comparatively close to the surface. Surrounding this central, copper zone area is a ring or zone of deposits that represents an ore zone intermediate between the copper and the zinc zone. Surrounding these deposits are others which seem to represent a combination of the zinc and lead zones and carry silver. The ring mentioned is far from circular and its width varies greatly from point to point. Furthermore, there are no sharp boundaries between areas in which deposits representative of different ore zones outcrop.

Practically, the facts just outlined have one important application. If, in any mining district, deposits representative of a relatively deep ore zone are found in a certain part of the district, and these deposits are surrounded by rings or zones of deposits representative of shallower zones, and a shaft is sunk on one of these outlying deposits, it is apt, if sunk deep enough, to pass through the same sequence of ore as one would pass over if he went from the shaft to that part of the district where ore representative of the relatively deep zone occurs.

Examples.—In order to illustrate the practical value of the facts presented in this chapter, the following problems, with correct answers in each case, are presented. They are not hypothetical, but are all drawn from the experience of the writer.

Problem 1.—A prospect shows a quartz vein that contains pyrite which carries a small amount of copper. Will the copper content probably increase with depth?

Answer: No. The deposit is probably near the top of the iron-gold zone from above which the copper zone has been mostly removed by erosion.

Problem 2.—A mine opened to a depth of 600 feet contains galena which carries much silver, iron-free sphalerite, gray copper, and pyrite, with increasing amounts of chalcopyrite in the lower levels. What change in the character of the ore would you predict if tapped by a tunnel at the 1200-foot level?

Answer: This deposit is of a rather peculiar type in that, while it is largely in the lead zone, it shows a decided commingling with the zinc zone below and with the silver zone (as indicated by the presence of the gray copper) above. Whether any considerable change will be shown in a vertical distance of 600 feet is problematical, but if it does occur it will take the form of an increase in the proportion of copper and a probable decrease in the percentage of silver. Development actually showed a large increase in the proportion of copper and considerable decrease in the amount of silver in the ore. The gray copper, which was rich in silver, was almost absent where tapped by the tunnel.

Problem 3.—A deposit which contains about equal proportions of iron-free sphalerite and pyrite contains just enough gold to make it workable. The shaft makes lots of water, and a tunnel to tap the vein at a depth of 800 feet below the bottom of the shaft is recommended. Will good ore probably be exposed by this tunnel provided faulting has not occurred?

Answer: The deposit is evidently near the bottom of the zinc zone. As no arsenic or copper is present in the ore, it is probably underlaid by the iron-gold zone. If any change occurs between the bottom of the shaft and the tunnel level, the sphalerite should become less plentiful and the pyrite more plentiful. Assays show that the gold is included within the sphalerite. A diminution in the proportion of sphalerite will therefore mean a lowering of the grade of the ore. As even a slight decrease in the amount of gold present will make the ore unworkable, it is likely that the ore exposed in the face of the tunnel cannot be mined profitably unless the saving in the cost of mining due to the fact that the ore need not be hoisted will offset the possible decrease in the value of the ore. Against this saving must be set the cost of driving the tunnel, however. The tunnel, when driven, exposed ore which contained comparatively little sphalerite and so little gold that it could not be mined profitably.

CHAPTER III

SOME MYTHS RELATING TO LEACHING

Some False Ideas Stated

Many miners and prospectors adhere firmly to the belief that practically every outcrop has been largely or entirely leached of any valuable metal or metals that it may originally have contained, and they contend that richer ore is certain to be encountered at that depth at which leaching has not occurred. They know that surface water contains atmospheric gases, earth derived salts, and organic acids in solution, and is, in fact, a weak chemical reagent that dissolves or leaches many minerals out of veins or other types of ore deposits through which such water slowly percolates downward. When such surface waters attack sulphides, powerful solvents are formed. These miners err in two particulars, however: first, in supposing that all veins or similar structures originally contained valuable minerals, and second, in assuming that valuable minerals are always leached out when attacked by surface water.

Veins and Similar Structures Do Not Always Contain Valuable Minerals

It is, of course, evident that valuable minerals cannot be removed from an outcrop by leaching unless the outcrop originally contained such minerals, which is not by any means always the case. Beautiful veins with sharply defined walls and a filling of quartz or other valueless gangue minerals and, possibly, barren sulphide (pyrite) are not unusual; replacements of limestone with silica or metamorphosed masses of limestone changed to garnet and epidote occur without the development of any important copper ores; fine shear zones show profound evidences of mineralization yet contain little or no valuable metal. We might extend this list to include all the structures known sometimes to contain ore. All may be, and frequently are, practically barren of valuable metals, yet their outcrops may show evidences of great leaching. It follows, then, that a highly leached outcrop is alone no very satisfactory indication that more valuable ore will be found below the depth to which leaching has extended. Other features outlined later must be carefully considered before venturing to make predictions of this kind.

Some Minerals Are Not Normally Subject to Leaching

That surface water constitutes an extremely effective leaching reagent is undeniably true. Weak as it is, there are few minerals

23

that are not affected by it when exposed to its action through vast stretches of time. Most of even the hardest and most siliceous rocks ultimately yield to its slow attack, soften, and disintegrate into residual soil; while great thicknesses of the more soluble earth materials, such as limestone, are dissolved and carried off in the surface drainage. That surface water does descend along fissures or other lines of weakness, including veins, is also a fact. It is equally true that many valuable minerals are attacked and changed by such water, and may often be more or less completely leached out of outcrops; **but all minerals are not subject to such changes, and, if changed, leaching does not necessarily occur.**

Lead sulphide (galena), for instance, alters readily to lead carbonate (cerussite), lead sulphate (anglesite), and other compounds when exposed to such water, but lead minerals are decidedly insoluble in natural solutions and considerable amounts of lead are very rarely transported downward in solution. What leaching does occur is usually almost entirely confined to the actual surface exposure of the deposit, and, at a few feet below this point, the ore is apt to be as rich as at any greater depth unless a shoot of high-grade ore occurs at depth. In fact lead ore found within a few feet of the surface is usually richer than the deeper ore, since it represents both the metal originally in the deposit at that point and the small portion leached from the actual surface outcrop, carried downward, and deposited a few feet below. This rich ore constitutes a secondary enrichment, but the difference in the grade of such ore and that underlying is not usually great.

Gold is another metal that leaching affects little if at all (excepting under peculiar conditions discussed in the next chapter). While it would be incorrect to say that gold is entirely insoluble in surface water, the amount dissolved in most deposits is inappreciable; and in the majority of instances it would be folly to hope for an increase in the grade of gold ore below highly leached outcrops.

Other metals rarely or never removed by leaching near the surface (excepting, perhaps, directly at the outcrop where both solution and wind and water erosion are effective) are tin, tungsten, molybdenum (in the form of massive molybdenite), chromium, aluminum, and iron. Where these elements were originally deposited as sulphides, percolating surface waters usually convert them to oxides or so-called "oxidized" compounds, but this does not involve the loss of any of the metals excepting in the case of sulphides of iron from which leaching may remove much of the iron in the form of sulphates.

Among the more or less valuable minerals not classed as ores that are not subject to leaching by surface water are most gem minerals (including diamond, emerald, sapphire, topaz, amethyst, garnet, etc.), barite, white mica, graphite, talc, and asbestos. It is true, however, that some of these minerals may be cracked by shrinkage and expansion of the outcrop when subjected to temperature changes or by the hydration and swelling of some of the minerals with which they are associated, and that some minerals such as precious topaz and amethyst often lose their color when exposed to the light. It is also a fact that asbestos in outcrops is often hard and brittle, although soft and flexible at greater depth.

It is, then, quite possible that a better quality of some minerals exposed in outcrops will be found at some depth, but such improvement in quality is due to differences in physical features, and not to the fact that the minerals themselves, or valuable metals contained in them, have been leached out of the outcrops.

It must be admitted that the preceding statements are general ones, and are not universally applicable. Deposits of easily soluble minerals may remain unleached in arid regions while relatively insoluble compounds may be removed by leaching in hot, moist climates. It should also be remembered that the nature of surface water will vary with the character of the rocks, soil, and vegetation over or through which it flows; and, this being true, it is not surprising that minerals normally quite insoluble are occasionally leached out of outcrops. This is true, for instance, of ores of tungsten. Whereas these minerals normally outcrop in a decidedly unleached and unaltered condition, at least in Arizona, reliable observers have reported a few instances of deposits in which ores of this metal have undoubtedly been dissolved by surface water, and have thus largely disappeared from the outcrops.

Cases such as that just mentioned are exceptions to the general rule, however, and no one is justified in sinking on a deposit of unmarketable material in the hope that any considerable proportion of relatively insoluble metals or minerals has been removed from the outcrop by leaching, and that, consequently, there is a probability of encountering a profitable deposit of these substances below the point to which leaching has extended.

Leaching or Mechanical Removal of Valueless Minerals May Result in Enrichment

If the valueless gangue material in a deposit that contains relatively insoluble, valuable minerals is of such a nature as may be readily dissolved or easily softened or disintegrated by weathering, the gangue may be carried off in solution, be washed away mechanically in the surface drainage, or be blown elsewhere by the wind. If the valuable insoluble minerals thus released are too heavy to be easily washed or blown away, the wearing down of the outcrop may result in the accumulation of an increasing amount of this heavy material on top of or close to the outcrop. Gold, diamonds, sapphires, peridots, and garnets are among the minerals that have been released from their matrices and have accumulated on the earth's surface in this way.

If the valuable heavy material is in a finely divided condition, it may be carried down mechanically into cracks extending downward from the outcrop of the deposit. This downward movement has occurred in some deposits of gold that are rich at "grass roots" and for a few feet below the surface, but are far less profitable, and are in some instances quite unworkable, at greater depth.

Where outcrops of deposits such as those just described occur on hillsides, gravity supplements the effects of running water, and the valueless decomposition or disintegration products of the gangue and the valuable minerals released therefrom are both apt to be carried down the slope to the gulches or valleys. There the gentler slope over which the water runs decreases its velocity and effectiveness as a transportive agent, with the result that the heavy minerals tend to accumulate in the gulches or valleys as placer deposits. Such deposits are, then, merely a form of enrichment that may in some cases result indirectly from leaching.

A quite different and much more important type of enrichment occurs when valueless gangue minerals are leached out of an ore deposit without the removal of valuable metals contained therein. Assume, for instance, that the material as originally deposited consisted of 30 per cent lead sulphide (galena), 40 per cent zinc sulphide (sphalerite or blende), and 30 per cent of insoluble gangue minerals. When such an ore is subjected to leaching, all the zinc may be removed, and the resulting material may contain 52.8 per cent of lead carbonate (cerussite) and 47.2 per cent of gangue minerals. Whereas the original deposit contained 26 per cent of metallic lead, the ore after leaching will carry 40.9 per cent of this element. Not only is the percentage of lead actually increased by this process, but the rich ore is marketable as mined, whereas the lead and zinc minerals would have to be separated before the unleached ore could be sold; also, the carbonate ore is more valuable than the sulphide since there is a saving in the fuel required to expel the sulphur in the unleached ore.

The process just outlined is illustrated by an even commoner and more important type of leached deposit in which the material has been so changed that the valuable metal can be extracted therefrom more easily and cheaply than would otherwise be possible. Deposits of this kind originally consisted principally of sulphides of iron (or of iron and copper) which contained gold and were associated with varying proportions of gangue. In the unleached condition such ore may be too low-grade to be shipped and smelted profitably unless first concentrated, and concentration may be too expensive for the ore to yield satisfactory returns. If, however, the material is attacked by downward percolating surface water, some of the iron, most of the copper, and practically all of the sulphur may be carried away in solution, leaving the residual ore in what is known as a "free-milling" condition; that is, the gold may be recovered therefrom by amalgamation, a process to which sulphide ores are not amenable.

Most of the surface deposits of so-called "rotten" or "rusty" quartz that contains free gold have been formed in this way, and the gold deposits that prospectors have always sought so eagerly, which have been worked so profitably on a small scale, have usually been of this or the placer type. In most instances these pioneers desert their workings when the free-milling ore has been exhausted and the unaltered primary sulphides below have been encountered. Sometimes whole camps, once thriving, have been abandoned when the free-milling ore has been exhausted, but, in other instances, men with sufficient capital to mine and concentrate or otherwise treat the sulphide ore have acquired the properties and developed the mines profitably and to great depths. Well-informed mining men believe that, in camps with names now all but forgotten, many lodes from which the oxidized outcrops have been removed by pioneers await the coming of men familiar with modern processes of ore treatment who have the means to put these processes into operation; and, as the frontiers disappear before the onward march of civilization, it is likely that increasing numbers of these old mines will be profitably reopened, and will again become important sources of gold.

CHAPTER IV

SOME MYTHS RELATING TO SECONDARY ENRICHMENT

Secondary Enrichment Defined

The term secondary enrichment is applicable in its widest sense to that portion of an ore deposit that contains ore of higher grade than was originally deposited there. At such a point the original or primary ore has later been enriched in some way. As the processes that form ore bodies in the first place are called "primary" ones, those processes which later change it are considered to be of a "secondary" nature. When such secondary processes lead to the enrichment of the primary ore, the latter is said to be secondarily enriched, and the resulting relatively richer mass of ore constitutes a secondary enrichment. If a secondary enrichment is composed principally of so-called sulphide ore, it is called a secondary sulphide enrichment.

Enriching Processes

In the preceding chapter it was shown how leaching or mechanical removal of valueless minerals from that part of a deposit above the ground-water level may raise the grade of the ore there. This increase in value is certainly the result of secondary agencies, and such ore is secondarily enriched; but this result has been brought about by the subtraction of valueless material from the mass, not by the addition of valuable minerals to the ore already in existence. This chapter will be devoted to the consideration of enrichments by the last mentioned process, and all seconary enrichments mentioned subsequently may be assumed to have originated in this way.

It was stated in Chapter I that surface water will attack some minerals and form powerful solvents that dissolve some or all of the metals in many primary ore minerals. The solutions so formed may work downward through the resulting more or less leached material until conditions are encountered that cause these metals to deposit as secondary ore minerals. In regions of relatively heavy rainfall this deposition ordinarily occurs at or near the ground-water level—that depth below which, for indefinite, but often great distances the earth is permanently saturated with water. Ores of copper are especially apt to be thus dissolved and redeposited, but secondary silver and zinc deposits are not uncommon and other metals more rarely undergo concentration in this way. Secondary deposits of this kind, beneath leached material, are occasionally very rich—of much higher grade than the primary ore from which they were derived. Deposits formed or enriched by descending water are termed supergene, while ores deposited from ascending solutions or gases are termed hypogene.

The preceding paragraph is a brief and highly generalized outline of the process under consideration. Since many primary ore minerals are of the type called "sulphides" (compounds of sulphur, arsenic, antimony, tellurium, bismuth, selenium, etc., with metals), it may be profitable to consider more in detail the processes that lead to the formation of secondary enrichments of gold and sulphide ores.

The Secondary Enrichment of Gold

As far as is known, gold is only taken into solution, carried downward, and redeposited when (1) manganic oxide (pyrolusite or psilomelane) is present in the ore body, (2) the descending water contains some chloride and is acid, and (3) carbonate minerals (calcite or limestone, etc.) are absent from the ore body and the rock adjacent thereto.

Under the conditions just outlined free chlorine is liberated by reaction between the chloride and manganic oxide and dissolves the gold. The metal would be immediately reprecipitated by the ferrous sulphate formed by oxidation of pyrite were it not for the fact that this ferrous sulphate is, on contact with manganic oxide, immediately converted to ferric sulphate which does not precipitate gold from solution. Thus, manganic oxide not only releases chlorine, the solvent for gold, but eliminates the compound that would otherwise at once precipitate it from solution. The gold taken into solution in the manner suggested may percolate downward to a point where the solution loses acidity through kaolinization of feldspar or other reactions; and this loss of acidity, together with the increase in the amount of ferrous sulphate that exists as the supply of oxygen grows scantier, causes the redeposition of the gold as native metal. This deposition will normally take place at or near ground-water level in regions of relatively heavy rainfall. As the outcrop of the ore body is removed by erosion, the ground-water level is depressed, so a secondary enrichment of the type now under consideration may contain not only much of the gold originally deposited between such enrichment and the present surface of the ground, but it may also be a repository for the precious metal once a constituent of that portion of the deposit long since obliterated by erosion.

Nearly all water that falls on the earth's surface and percolates downwards through ore deposits contains an appreciable amount of some chloride—usually sodium chloride or common salt. Further, pyrite is an almost universal constituent of mineral deposits, and its oxidization imparts acidity to such water. These two factors in the enrichment of gold (the presence of chlorides and acidity) may, therefore, be assumed to exist in practically all deposits. It may be said, however, that almost complete absence of iron (in the form of the yellow or brown hydrated oxide called limonite, or the red oxide called hematite) from the leached ore is an unfavorable feature since it indicates that there was little or no pyrite in the original material.

The upper, leached portion of deposits in which enrichment of gold has taken place will, then, consist of gangue (often very siliceous), limonite or hematite, and psilomelane or pyrolusite (usually the former), but all the gold is rarely removed from this material. At no place will the gold content of the oxidized ore be high, however, and the residue of this metal will be distributed irregularly through the deposit in patches or pockets. The gangue will usually be decidedly fractured or porous. A compact, unfractured quartz gangue is almost never found in deposits that have yielded gold enrichments below the zone of leaching.

Deposits such as those under consideration have occasionally been worked above the water level as a source of manganese ore, but ordinarily the proportion of manganese dioxide is too low and the percentage of silica is too high to make such ore marketable as mined.

When a gold enrichment is encountered through downward extension of the workings, the primary ore will be found to consist of pyrite or some other sulphide that contains iron and pink carbonate of manganese (rhodochrosite or manganese spar), pink silicate of manganese (rhodonite), or, more rarely, black sulphide of manganese (alabandite). The gangue will probably include quartz and possibly other minerals, of which calcium fluoride (fluorite or fluor spar) is the most common. Sulphides other than pyrite may also be present. Fractures and other cavities in the primary ore will be filled or coated with rich manganiferous gold ore, kaolin, and limonite. Increased silver, copper, and zinc values may also be encountered if primary metals of these minerals have also been subjected to leaching. Oxides and suphides are apt to be closely associated in the enrichment, especially in the upper part, and the grade of the ore will be found to decrease rather suddenly as the enrichment is penetrated and the unenriched primary ore below is encountered. This primary ore will be gold bearing, but its grade will often prove so low as to make the mining of it unprofitable. The primary gold may be mechanically included in the sulphides or may be free in the gangue.

While no reputable geologist would venture to state positively that gold enrichments are never formed except under the conditions outlined, it can be said that no valuable enriched deposits of this metal appear to have originated in any other way; and it would certainly be folly to sink on any gold deposit of less than commercial grade in the hope of striking a secondary enrichment of gold below unless the leached portion of such a deposit shows the characteristics described. Even then, the operation must be regarded as speculative.

The Secondary Enrichment of Sulphides

Where primary ore consists of sulphides, downward percolating surface water will develop sulphuric acid and iron sulphate which act as solvents for the other sulphides. The metals thus dissolved may be carried downward until reprecipitated by the influence of unchanged primary sulphides, decrease of acidity of the solution, or other factors. Such deposition usually occurs at, or just below, the permanent ground-water level, underneath the zone of leaching; but, as is discussed later, when the ground-water level lies at great depth, all the deposition may take place in the leached zone, although at a considerable distance below the surface.

Where copper is deposited above the ground-water level, it occurs as green or blue hydrous carbonates (malachite and azurite), blue hydrous silicate (chrysocolla), red or black oxide (cuprite and tenorite), or native metal; while, where precipitated below the water, it is present as secondary sulphides, such as chalcocite (copper sulphide or glance) and covellite (copper sulphide), associated with the primary ore minerals and gangue.

Silver usually precipitates as chloride, bromide, or iodide (horn silver) or as native silver above the ground-water level, and as silver sulphide (argentite or silver glance) or sulpho salts of silver, such as the sulphantimonites (stephanite or brittle silver and pyrargyrite), the sulphantimonite containing copper (polybasite), or the sulpharsenite (proustite or ruby silver) when deposited below water. These minerals are not infrequently also commingled with the native and horn silver above the ground-water level.

Zinc precipitates as carbonate (smithsonite), hydrous carbonate (hydrozincite), and silicate (calamine) above the ground-water level, and as sulphide (wurtzite or sphalerite) below the water level. The sulphide is rarely deposited as a secondary mineral, however.

Lead deposits as carbonate (cerussite), sulphate (anglesite), phosphate (pyromorphite), arsenate (mimetite), vanadate (vanadinite), molybdate (wulfenite), oxides (minium and massicot), plumbojarosite (hydrous sulphate of lead and iron), etc. above the ground-water level, and is very rarely carried below the water.

Where all conditions are normal and the ground-water level is comparatively close to the surface, four more or less clearly developed and sharply differentiated zones of varying thickness may be formed. These zones may be thus designated:

1. The leached or low-grade oxidized zone
2. The secondarily enriched oxidized zone
3. The secondarily enriched sulphide zone
4. The zone of primary, low-grade sulphides

Zones 1 and 2 are formed above the ground-water level, zone 3, normally at or just below this level, and zone 4 underlies zone 3. These zones may be sharply differentiated, but are more apt to be transitional one into another. The most sudden and distinct change is usually found between zones 2 and 3 (or 1 and 3 where 2 is lacking). Overlying zones may penetrate deeply into the zones below, especially along water courses; and the three upper zones, even when sharply defined, are usually decidedly warped masses of varying thickness, not tablets with plane upper and lower surfaces.

The thickness of zone 1 depends almost entirely on the depth of the water level; if the water lies far below the surface, this zone will be thick. Zone 2 is usually thin or may even be entirely missing if the water level lies at shallow depth. It is sometimes very thick when permanent water is not encountered until great depths are reached, and may then contain secondary sulphide minerals as well as oxidized ores.

Zone 3, when present, is usually the thinnest of all; its thickness can ordinarily be stated in tens or scores of feet.

The eroded portions of a deposit may have contributed materially to the valuable metal in zones 2 or 3, just as was true of gold enrichments.

The leached or low-grade oxidized zone includes the leached surface outcrop. It may be composed of nothing but quartz or other gangue minerals, which, when susceptible to alteration, are often hydrated and softened; but it is usually more or less stained with oxides of iron, carbonates or the silicate of copper, and, less frequently, oxide of manganese. Frequently the percentage of iron is high when important enrichments of copper occur below, and such highly ferruginous outcrops form the "iron hats" or iron cappings so eagerly sought by prospectors when copper was in demand. Where a deposit consisted originally chiefly of silver or zinc ores, they may be almost or entirely removed from this zone; and, even when present, they are not represented by highly colored stains. Lead deposits are frequently exceptionally rich in the uppermost zone, and the oxidized ores of lead often run higher in silver than did the primary ore. In this zone may also be found free gold as well as unchanged kernels or residual masses of primary sulphides. These primary sulphides may be coated with secondary sulphides and, still further from their centers, with oxides. Sinking on such deposits will often reveal the same sequence of minerals as is shown when passing from the exterior to the interior of these residual masses.

Among the features that may be shown by this zone, which are indicative of the presence of enrichment at depth, are:

1. A large proportion of iron oxide.
2. (In cases of copper enrichment). Plentiful green or blue copper stains and residual masses of primary copper or copper-iron sulphides.
3. Profound fracturing or decided porosity produced in any other way, such as by the complete removal of sulphides.

4. Thorough softening through kaolinization of the vein matter or wall rock.

5. Absence of carbonates in gangue or walls.

Some of the disseminated copper ores of the Southwest have decidedly silicified or bleached and kaolinized outcrops, and highly siliceous surface crusts occur. These crusts are usually composed of a mixture of chalcedonic quartz, copper silicate, and copper carbonates. The outcrop, or the material just below if the siliceous cap is present, will show residual specks or patches of limonite or hematite (red oxide of iron). These specks or patches may be of very irregular shape or may have the form of pyrite crystals. Some empty cavities may be found, especially very near the surface. If the specks or patches of iron oxide all have the form of pyrite crystals, the chances of finding a valuable copper enrichment below are not so good as when a large proportion of them are of irregular shape. Numerous, closely spaced veinlets of quartz or iron oxide are a favorable outcrop feature. This is particularly true if the veinlets are composed of pure quartz.

The secondarily enriched oxidized zone may be produced by the oxidization of an enriched sulphide zone that is settling as a result of rapid erosion of the surface and consequent depression of the ground-water level, or, when the ground-water level lies very deep, this zone may be formed through direct deposition of the metals as oxides from the downward percolating solutions. Where it has originated as first suggested, it will be relatively thin, but rich; and, if the other explanation is applicable, it will be comparatively thick but of lower grade. In the former case a considerable proportion of primary sulphides may occur with the oxidized minerals which will occupy cracks or vugs in the primary minerals. Kaolin, chalcedony, opal, and other secondary gangue minerals are apt to be present in this zone.

The secondarily enriched sulphide zone is usually the "bonanza" zone, and is eagerly sought. Once found, however, it is certain that the values will decrease, often suddenly, at no great depth below. Primary and secondary sulphide minerals will be intermingled in this zone, and soft, pulverulent masses of mixed sulphides are common. Some secondary oxides and gangue minerals may be present, especially in the upper part.

The zone of primary, low grade sulphides contains the unchanged primary ore and gangue minerals. The material is usually compact and the country rock is fresh. Hematite is the only oxide commonly found associated here with the sulphides. The grade of the ore is frequently, although not necessarily, too low to permit its profitable extraction.

When applying the facts already outlined it should be borne in mind that a change from lean to good ore may be due to the appearance of an ore shoot and may be entirely unrelated to secondary enrichment.

Some Fallacious Ideas.

The phrases "secondary enrichment" and "secondary sulphide enrichment" are comparatively recent additions to the miner's vocabulary; in fact, they were rarely heard twenty-five years ago, and practically all of the great mass of data relating to enrichments and the enriching processes involved that is now available has been gathered during the last three decades. Possibly it is because the ideas involved are comparatively new and not thoroughly understood that prospectors and miners are prone to believe that enrichments are much more common than experience has shown them to be. The processes outlined in this article have certainly been extremely effective in the formation of some very valuable ore deposits, and it is perhaps only natural, since hope is the father of belief, that many men should confidently expect to find such enrichments beneath practically all oxidized outcrops. Those who have had the opportunity to observe and study ore deposits at all exhaustively, however, do not hesitate to doubt the validity of the following statements:

1. Secondary enrichment should be expected below *every* oxidized outcrop.
2. If valuable metal has been leached from the outcrop, it will always be found *below* as an enriched deposit.
3. Where sulphides have been subject to such leaching, secondary enrichments found below will always be in the form of *sulphides*, that is, secondary sulphide enrichments will always be formed in such cases.
4. An enrichment, if formed, will always be found *vertically* below the leached material from which the valuable metals in such enrichment were carried downward.
5. Secondary enrichments, when formed, will always be found at or just below the ground-water level.
6. If water is encountered when sinking and no enrichment is found directly below this water, it is always useless to sink further.

These fallacies will be discussed briefly in the order mentioned.

1. No enrichment can be formed if the primary material did not originally contain a valuable metal or metals. Profound oxidization may occur, for instance, in deposits of barren iron sulphide.

2. It is undeniably true that solutions of valuable metals, while percolating downward, may encounter strong upward flows of water before depositing their metallic content, and may work their way to the surface, where the metal may be dissipated.

3. It has been shown that, where the ground-water level lies at great depths, as in semiarid regions such as Arizona, deposition may occur through a relatively great range of depth, and that comparatively low-grade oxidized deposits may then be formed.

4. Where an ore deposit occurs in well-bedded rocks with a more or less pronounced dip (and even under other conditions), the solutions may work downward and *laterally* along the bedding planes and other channels, and their metallic content may ultimately be deposited far to one side of vertical lines extending downward from the points where the metals were dissolved.

5. Secondarily enriched oxidized deposits above the ground-water level are by no means uncommon in semiarid regions. It should also be remembered that sudden or decided elevations or depresions of the ground-water level sometimes occur. If such changes have occurred recently a zone of secondarily enriched sulphides may exist above or below the present ground-water level.

6. Not only will fluctuations in the ground-water level leave secondary enrichments below the upper surface of the water, but it is also true that some water is often encountered overlying relatively impervious beds at one or more horizons above the true ground-water level. Such "perched" water levels may have little influence upon descending solutions.

Important Influence of Carbonates

Well trained, experienced mining engineers have not infrequently overlooked the importance of one factor when they have attempted to decide whether secondary enrichment has or has not probably taken place. It is not surprising, therefore, that many prospectors should have made the same mistake, yet there is really little excuse for anyone to do so since the significant features are easily recognized. The factor involved is the presence or absence of carbonates in the gangue or walls.

Descending solutions that dissolve metals are acid and no metals will be dissolved unless the solutions are acid. Any condition, therefore, that neutralizes the acidity of the solutions will prevent the dissolving of most metals. There can then be no transportation of these metals to some other position and no enrichment can occur.

Calcium carbonate is a substance that will quickly neutralize any acid solution that comes in contact with it. This substance, as the mineral called calcite, very frequently occurs in gangue material, and, as limestone, it constitutes a common rock. Although any mineralogist can readily recognize calcite by its physical properties, it and limestone may be most easily and certainly identified by the fact that, if placed in cold hydrochloric (muriatic) acid, they will dissolve with vigorous effervescence (bubbling).

If any considerable amount of calcite is present in the gangue of a sulphide ore body through which surface water is descending, any metal excepting silver that may be dissolved will be transported only far enough to bring the solution containing it into contact with calcite. The metal will then immediately be deposited. There can, under such conditions, be no considerable downward migration of the metal. In fact, if calcite is plentiful in the gangue, a metal other than silver cannot be transported to any appreciable extent. It will oxidize and remain at or very close to the point where it was deposited as primary ore. Silver may be transported in spite of the presence of carbonates if no chlorine or other halogen element is present.

Calcium carbonate is not the only natural substance that neutralizes acid solutions. Any carbonate will have this effect, but carbonates other than calcite are relatively uncommon constituents of primary ore bodies, although ankerite (calcium-iron-magnesium carbonate) and dolomite (calcium-magnesium carbonate) are plentiful in some deposits. Feldspar also acts as a neutralizer, but its action is much slower than that of calcite, and, unless the solutions are very weakly acid and are percolating down through feldspar with unusual slowness, the migration of metals through a gangue that contains feldspar may occur and an enrichment may form.

Not only does the presence of considerable calcite in the gangue of an ore body prevent the migration of metals in solution, but it is undeniably true that limestone walls may also have the same effect. Whether limestone will or will not act as a neutralizer of acid solutions and a precipitant of the dissolved metals depends wholly upon whether the solutions come in contact with it. If a vein contains no calcite and it is sufficiently porous so that descending solutions work downward through the vein material without coming in contact with limestone walls, an enrichment may be formed below, but such conditions are unusual. In most instances most of the downward movement of solutions is along one or both walls and then the solutions will be quickly neutralized if the walls are of limestone.

There is a possibility that acid solutions that descend along limestone walls may sometimes attack and react with the walls and form protective layers of alteration products which will themselves not neutralize acid solutions, and that thereafter migration of metals may take place. If the inclination and physical condition of a vein that contains no calcite is such as to indicate that descending solutions were concentrated along one or both walls of limestone, yet the wall material does not effervesce with hydrochloric acid and other conditions indicate that leaching of one or more metals has occurred, there is a chance that a secondary enrichment may exist at greater depth, but in general one cannot hope to find enrichment of deposits in limestone. Similarly, there is a bare possibility that deposits that contain some calcite in the gangue may be leached of one or more metals, to some extent, if the calcite is not very plentiful or is so situated in the deposits that a considerable part of the solutions does not come in contact with it and descends along walls that are not limestone. It is rarely safe to predict the existence of a secondary enrichment when there is any calcite present in the gangue, however.

CHAPTER V

THE PHONOLITE MYTH

The Title Explained

Three or more decades ago, when Cripple Creek, Colorado, was in the heyday of its glory, it became widely known that the rich deposits of telluride of gold found in that camp were associated with a peculiar rock that the scientists classified as phonolite. This rock as found at Cripple Creek is extremely dense, has a somewhat greasy luster, occurs in various tints of dull green, gray, and brown, and shows few easily identifiable minerals other than small, scanty crystals of glassy feldspar and, sometimes. numerous very tiny, slender black crystals of aegerite pyroxene. Thin slabs, when suspended or held in the proper way and struck with a hammer or pick, ring like a bell, hence the name, from two Greek words meaning "sound" or "tone" and "stone."

As a late manifestation of the volcanic activity that gave rise to the ancient Cripple Creek volcano of which the only present visible evidence is the crater filled with volcanic tuff and breccia (cemented fragmental material), this phonolite broke as molten lava up through the crater filling,. and quickly solidfied in the form of dikes. Dikes of syenite and of a dark colored rock, locally termed basalt, were formed in a similar fashion; and, finally, there arose through the cooling, shrinking, and cracking rock that filled the crater (and to a lesser extent, through the granite that formed the crater walls) the solutions from which the rich gold minerals were deposited. It is extremely doubtful if the phonolite had anything to do with the deposition of the gold from these solutions. In fact the association of the two substances was probably wholly accidental. Gold is by no means always associated with phonolite dikes at Cripple Creek; rich deposits have been found in the granite, the basalt, and the breccia which no miner would call phonolite although it is largely made up of phonolitic fragments. Yet gold and phonolite have so often been found together in the camp that Cripple Creek miners have come to regard phonolite as a rock almost certain to contain gold; and these men, moving to other camps, carried the fallacy with them and established it in the minds of others; so, in a surprisingly short time, it has become a widely accepted theory. That the idea under consideration is wholly fallacious is undeniable; indeed, it is probably true that very few other rocks are less likely to contain gold than is phonolite. The presence of

37

this rock should, then, probably be regarded by prospectors as an unfavorable rather than a favorable feature; but for a generation or more any rock that has the general appearance of phonolite has been eagerly sought. If a vein or a shear zone in such a rock were found, the discoverer at once assumed that his fortune was made; and tens and hundreds of thousands of dollars have been wasted on such prospects. As a matter of fact the rock thus located has seldom been phonolite even though some of it may have "rung like a bell." Phonolite can rarely be recognized with certainty by other means than a microscopic examination of a thin section; but, assuming that some prospects contained true phonolite, there has never been any valid reason for their development unless other evidences of mineralization were present.

The misconception just discussed constitutes the "phonolite myth," but it seems desirable to extend the term so that it covers all theories of the association of metals with certain types of rocks, provided such ideas are based on conditions observed in only one or two isolated occurrences.

Too strong emphasis cannot be placed upon the statement that, even though ores of a certain metal are associated with a particular species of rock in one mine or locality, there is not necessarily any valid reason for expecting a similar association elsewhere. Until one has had an opportunity to study occurrences of such ores in scores or even hundreds of places, he is in no position to generalize in this way. An enormous sum in the aggregate has doubtless been wasted by men who have attempted to make such generalizations without adequate knowledge of the facts.

The Rocks Associated with Gold Deposits

The old saying that gold is where you find it is so true that it is unsafe to claim that deposits of this metal will not be found in any rock. It is, however, equally unwise to regard a specific kind of rock as being especially apt to contain important deposits of gold.

In general it may be said that a series of extensive flows or intrusions of igneous rocks of Tertiary or later age should be carefully prospected for this metal, especially if these rocks are relatively acid (light in color and weight), such as rhyolite, andesite, acid porphyry, or related species. The presence of some basic (relatively dark and heavy) rocks, such as basalt, diabase, and basic porphyry, or of associated sediments need not be regarded as especially favorable or unfavorable features.

Great masses of acid plutonic rocks (granite, syenite, diorite, and related species) are not usually regarded by prospectors with as much favor as are those other acid igneous rocks mentioned in the preceding paragraph, and this statement is particularly true as regards granite. Although valuable deposits of gold have been found in granite, they are decidedly exceptional, and the values are not apt to persist downward to any considerable depth. Valuable gold deposits have, however, been found around the borders of huge outcrops of granite. Gold deposits in the other plutonic species mentioned are apt to be also associated with dikes or other instrusions of nonplutonic igneous material.

Large masses of basic igneous rocks such as gabbro, peridotite, basic porphyry, diabase, and basalt rarely contain workable deposits of gold, and should generally be regarded with disfavor by prospectors for this metal.

Metamorphic rocks such as schist and gneiss need not be regarded either favorably or unfavorably; but, if such rocks contain dikes or other intrusions of igneous material, it may pay to prospect them.

Areas of sediments (sandstone, limestone, shale, conglomerate, etc.) and their metamorphosed forms (quartzite, marble, slate, etc.) unassociated with and far from igneous rocks rarely contain gold, or, indeed, ore of any other metal; although the zinc-lead deposits of the Mississippi valley and elsewhere are notable exceptions to this rule.

The foregoing statements do not, of course, apply to alluvial deposits; and no consideration is given to placer deposits in the rest of this chapter. Placer gold may be associated with boulders and smaller fragments of almost any kind of rock, but the gold eroded from deposits in Tertiary lavas is usually so finely divided that it is not easily concentrated in placers.

The Rocks Associated with Silver Deposits

Much that has been said about gold is equally applicable to silver ores, and, indeed, appreciable amounts of silver are almost always present in deposits of gold.

Important deposits of nearly gold-free silver-lead ores occur, however, in numerous localities as veins or replacements in limestone that is usually cut by dikes or intruded by sills (sheets of material that have forced their way while molten between layers of solid rocks that lie horizontally or nearly so) of igneous material, and areas showing such association of limestone and igneous rocks should always be prospected. It should not be assumed, however, from what has just been said that silver-lead ores are always found in association with the rocks mentioned.

The Rocks Associated with Lead and Zinc Deposits

Deposits of nearly silver-free lead-zinc ores, associated with limestone and chert, occur at several points in the upper Mississippi valley. These ores usually occur filling connecting vertical (or steeply dipping) and horizontal fissures (pitches and flats), but they are also found scattered through broken or porous limestone and as disseminations in this rock.

Outside of the Mississippi valley, lead-zinc ores, either with or without variable proportions of gold or silver, are found in both igneous and sedimentary rocks (especially limestone) that are cut by dykes or other intrusions of igneous rocks, but productive deposits of lead are rare in igneous rocks.

Gold-free zinc ores are found in limestone in the Mississippi valley; elsewhere they appear most frequently in limestone that contains dikes or other intrusions of igneous rocks; while gold-bearing zinc ores are probably most likely to be found in close association with igneous rocks.

The Rocks Associated with Copper Deposits

Copper ores are often found in limestone closely associated with igneous rock that is neither extremely acid nor basic, or in igneous rocks such as monzonite or allied species. The limestone has often been more or less converted to brown or green garnet or green epidote, but garnetization and epidotization of limestone is not a certain indication of the presence of workable deposits of copper ores.

, Another frequent associate, especially in Arizona, is schist. In this state the quartz-muscovite (white mica) Yavapai and Pinal schists have been very productive at several points.

Copper is also found associated with almost as great a variety of other rocks as is gold or silver. Important disseminated deposits occur in "porphyry."

The Rocks Associated with Mercury Deposits

Although mercury ores have been found in many different rocks, they are usually not far from acid igneous rocks. The deposits may, however, be actually located in other material, as at New Almanden, California, where the ore lies between serpentine and shale, although a rhyolite dike lies close to, and roughly parallels, the ore body.

The Rocks Associated with Chromium Deposits

It is possible to say very definitely that chromium ore is very rarely found associated with anything but large, basic, plutonic masses or their altered equivalents, especially peridotite and the relatively soft, greasy feeling alteration product called serpentine; in fact, serpentine and peridotite areas should always be prospected for chromite (chromium-iron oxide), the only important chromium ore, and it is almost certain to prove a waste of time to seek this mineral in other rocks.

The Rocks Associated with Tin Deposits

Tin is another element the ores of which show a pronounced tendency to occur in association with one or two particular types of rock, namely, pegmatite or greisen (quartz with light colored mica). Tin-bearing masses of these rocks around granitic areas are found in a number of localities, but such deposits are rarely of commercial importance, since a large part of the world's tin output comes from placers.

The Rocks Associated with Tungsten Deposits

The ores of tungsten are generally found in pegmatite dikes or in veins that are commonly associated with granitic rocks or acid porphyry.

The Rocks Associated with Platinum Deposits

Although some platinum occurs as rare platinum minerals in deposits mined mainly for their copper content, which have the usual rock associates of copper ores, and has been recovered profitably from very basic dikes, almost all of the platinum produced is obtained from placer deposits of the native metal alloyed with more or less iron and with other members of the platinum group of metals. When placer platinum is traced back to its source it is almost invariably found to have come from some variety of the very basic, plutonic igneous rock called peridotite, or its alteration product, serpentine. The streams draining large areas of these rocks should always be panned for platinum.

Conclusion

The list of metals already briefly discussed might be extended somewhat if it would serve any good purpose to do so. In most instances, however, it would be found that in each case the associated rocks vary widely in different localities—the metal under consideration shows no special tendency to be deposited in or close to any particular kind of rock.

To be sure, it has been shown that the "phonolite myth" is not entirely a myth when applied to deposits of chromium, tin, metallic platinum, and tungsten. These metals or the minerals containing them are so nearly universally confined to distinct types of rocks that it should be regarded as a waste of time to search elsewhere for them. It should not be thought, however, that, because chromite, for instance, is almost never found in commercially important amounts in anything but peridotite or serpentine, areas consisting of these rocks are sure to contain deposits of chromite. The phonolite myth usually takes the form of some such idea as that just condemned, and, in this form, it is most positively fallacious. There are doubtless hundreds of masses of serpentine that contain no important deposits of chromite to one in which this material occurs in such amounts as to make profitable mining a possibility and, although peridotite and serpentine are the mother rocks of metallic platinum, such rocks are nowhere mined as sources of this metal except in one district in South Africa. It can be stated, further, that thousands of pegmatite dikes exist in which no appreciable quantities of tin and tungsten minerals are found.

It would be very gratifying if all that one had to do to discover an ore deposit was to find an outcrop of a certain kind of rock, for such outcrops are not usually difficult to locate, but, unfortunately, many factors other than the presence or absence of certain rocks enter into the formation of all ore deposits, and it would be the height of folly for a prospector to sink on any rock unless he has some other evidence of the possible existence of ore. Even when such evidence exists the presence of a certain kind of rock should not materially encourage the miner, except when he is seeking one of the metals that have been shown to be commonly associated with particular rock species.

Of course, what has been said is not always applicable in districts of limited size. Within a given area of greater or lesser extent, the ore deposits often show many points of similarity, and the veins found there may, in the majority of instances, yield good values when associated with a certain type of rock and prove barren or nearly so when this rock is absent; but, excepting in the case of certain metals mentioned, utterly different conditions may prevail in other districts, and too much reliance should not be placed, even in one district, upon the influence of rock associates unless experience has shown that a certain rock is almost invariably associated with good ore.

Most prospectors are seeking deposits of gold, silver, or copper, and these are the very metals with which the widest variety of rocks is associated. It is probable that, if three different mining geologists were asked to express an opinion as to what distinct species of rock is most frequently associated with valuable deposits of gold, three different answers would be made. While there is no one rock that may be regarded as a reliable indication of the presence of economically important occurrences of either gold, silver, copper, lead, or zinc ores, it is true that it is usually a waste of time to prospect an area for ores of these metals unless igneous rocks (especially those occurring in dikes, sills, and other intrusions of moderate size) are present. It should be remembered further that sediments, especially limestone, or schists cut by moderate-sized intrusions of igneous rocks, or great masses of recent volcanics are especially promising fields of effort. By studying geological maps prospectors can thus select the areas most likely to reward their efforts.

CHAPTER VI

SOME MYTHS RELATING TO OUTCROPS

Introductory

There is nothing with which prospectors are so directly and intimately concerned as outcrops. Only operators with much capital can afford to prospect ground with a diamond drill, or to seek blind veins by sinking on barren ground; all others must search for actual outcrops of ore bodies, and are quite naturally interested in and deeply concerned with the characteristics of any outcrops discovered.

It is unfortunate that individuals rarely prospect in more than a limited number of regions in each of which all the geological features are apt to be somewhat similar and that the number of outcrops actually developed by any one prospector is usually comparatively small. Each thing he sees and finds may have such a vital effect upon his interests that it makes a deep impression upon him, and he is apt to formulate laws and rules that, while they may apply to a few prospects or districts, are by no means universally true.

Under the conditions just suggested it is not surprising that many unwarranted generalizations have been made, and have been widely accepted. At any rate it is certainly true that a number of the current ideas relating to outcrops are fallacious, excepting, perhaps, when applied to relatively restricted areas, and this chapter will treat of some of these fallacies.

Valuable Deposits Not Always Marked by Prominent Outcrops

Only a tenderfoot or a mining-stock salesman would venture to make the assertion that the value of a deposit can be measured by the prominence of its outcrop. The wildcat mine promoter seems to be especially addicted to the myth that an outcrop that projects well above the surface of the ground and can readily be traced for thousands of feet is sure to bring riches to its possessor. Experienced prospectors know that the exact opposite is often the case.

Is it not true that some outcrops that project above the surface like walls or form ridges or mounds overlie important ore deposits? Unquestionably they do. W. A. Stratton, a carpenter from Colorado Springs, who knew little or nothing about minerals, located the Independence Mine at Cripple Creek on such an outcrop, and scores of similar discoveries could be cited. It is equally true, however, that the majority of very productive ore bodies have outcropped very inconspicuously; in fact the richest ore shoots have frequently been completely hidden beneath wash or soil and vegetation. At first sight the evidence appears decidedly contradictory, and the inexperienced prospector is not to be blamed if he thinks of the matter as one without rhyme or reason. But nature does not operate in this way; her effects are produced by the operation of fixed laws which it is the task of science to discover; and the law governing that phase of outcrops under consideration is simple. It may be stated thus: Any outcrop of an ore deposit will project noticeably above the surrounding earth materials if it is entirely or very largely composed of minerals that are harder than the enclosing rocks.

It should be remembered, however, that it is not merely a question of the relative hardness of the unaltered country rock and of the primary minerals in the ore deposit; weathering of the country rock and weathering or leaching of the ore or gangue minerals may affect their hardness profoundly, and it is the relative hardness of the alteration products of these materials that counts unless the area prospected is in an arid or semiarid region where weathering is a negligible factor.

Among the outcrops that are apt to be harder or more resistant to weathering than the surrounding rocks are the following:

1. Massive quartz veins with or without free gold, scanty sulphide minerals, etc. (unless badly crushed or sheared).

2. Pegmatite dikes (may contain tungsten, tin, gems, etc.).

3. Limestone-igneous rock contact deposits marked by the development of much garnet and epidote, such as frequently contain copper.

4. Any type of deposit that contains so much iron sulphide as to yield a high proportion of limonite (hydrous iron oxide) when weathered (the "iron hat" sought by prospectors for copper deposits). Such outcrops may also be valuable for the free-milling gold contained therein.

5. The siliceous crusts sometimes found over the leached outcrops of disseminated "prophyry" copper deposits. In fact igneous rocks that contain disseminated copper minerals are in general apt to be as hard as or harder than surrounding rocks, and to outcrop as hills which are sometimes high.

6. Chromite or magnetite deposits in serpentine.

Outcrops that are, on the other hand, likely to be softer or less resistant to weathering than the surrounding rocks are the following:

1. Shear zones and brecciated veins (zones of rock broken into angular fragments, with the cracks between the fragments filled with ore minerals and gangue) in which the fractures have been incompletely cemented by later deposition of minerals. A wide variety of metals may be found in such deposits as well as in those next mentioned.

2. Fahlbands (sulphide impregnated zones of schist).

3. Any kind of deposit in which the gangue is predominantly calcite (calc-spar or calcium carbonate), dolomite (magnesian spar or calcium-magnesium carbonate), fluorite (fluor spar or calcium fluoride), rhodochrosite (manganese spar or manganese carbonate), or barite (heavy spar or barium sulphate). Some of these minerals are soluble, some alter to substances of lesser volume, and all are relatively soft. Silver and lead ores are very frequently found in such gangues.

Although there are so many exceptions that the statement has little value, it may still be said that deposits of tungsten, tin, molybdenum, and copper minerals and of chromite and magnetite are apt to outcrop prominently as ridges or hills; that deposits of silver and lead are likely to have inconspicuous outcrops (such deposits may even be marked by a notch or a trench where they cross divides); and that outcrops of deposits of gold and zinc are in some places conspicuous, and elsewhere inconspicuous.

Outcrops of veins that stand up prominently at some places and are indistinct at other points should be prospected where most obscure, since there they may have contained unusually high proportions of ore minerals which commonly yield soft or soluble products on alteration.

Veins Do Not Always Widen Below Their Outcrops

Prospectors are prone to expect not only that relatively high-grade ore will always be found at some depth but also that narrow veins will widen below their outcrops. It has already been shown in the preceding chapters that leaching and secondary enrichment furnish some justification for the first mentioned idea, but there is rarely any real reason to expect that a vein will widen with increased depth. In fact experience has shown that veins are just as apt to pinch as to swell below their outcrops.

In general it may be said that veins that show considerable variation in width from point to point along their outcrops may be expected to exhibit similar variations in thickness when followed downward. It is, of course, equally true that a vein of approximately the same thickness wherever exposed on the surface will have about the same width at all points below the outcrop. Fault fissure veins, brecciated veins, shear zones, contact deposits, and any tabular type of deposit in connection with which impregnation or replacement of the wall rock has occurred will usually show considerable variation in width both horizontally and downward; while simple fissure veins and fahlbands often show comparatively little variation in width for considerable distances in all directions.

Where a vein outcrops in tough, compact, difficultly replaceable rock, and there is reason to believe that at some depth it intersects very brittle, porous, or easily replaceable material, it is likely that it will be wider in such rocks than at its outcrop. If the positions of the two classes of rocks just mentioned are transposed, it is probable that the vein will eventually become narrower rather than wider when followed downward.

If two veins gradually converge downward they may unite, or, of course, one may cut through the other. If they are approximately parallel and contain the same ore and gangue minerals, there is a good chance that they will unite, especially if the convergence is gradual. If more than two veins contain the same ore and gangue minerals, have nearly parallel outcrops, and all converge, they are fairly sure to unite, but such conditions are rarely found in deposits other than those of shallow type which have suffered little or no erosion since their formation. The width of a vein formed by the union of two or more converging veins is usually less than the sum of the widths of the veins that have united at depth.

The rock between a number of converging veins that unite at depth is often badly shattered and mineralized. The combined width of the outcrop of the converging veins and of the mineralized rock between may be a score or more of feet, and may be mistaken for the outcrop of a wide lode of complex character. Indications of the convergence of the component veins of such a mass should be sought before assuming that the conditions existing on the surface will persist with downward development.

No Fixed Relationship Exists Between Length of Outcrop and Depth to Which Development May Profitably be Carried

Prospectors occasionally express the conviction that a vein or other tabular form of ore deposit will extend downward a distance at least equal to the length of the outcrop. If they are thinking of the extent of the vein regardless of the grade of ore contained therein, or of the horizontal and downward dimensions of ore shoots (to be discussed in a forthcoming chapter), the idea stated cannot be branded fallacious. In most instances, however, the dimensions under consideration are (1) the total length of the outcrop of a vein or similar type of deposit, including both barren and mineralized portions, and (2) the depth to which the grade of any ore found therein will remain sufficiently high to permit profitable mining.

The idea that any reliance can be placed on the rule that ore can be mined profitably to a depth equal to the length of the entire outcrop is without foundation, and is as fallacious as is the notion that the two dimensions under consideration bear in general any fixed numerical relationship to each other.

The length of the main or more productive portion of the Comstock Lode of Nevada is about 2 miles, while the length of the entire deposit, including the narrowed and split ends, is approximately 4 miles, or about 21,000 feet. The lode pinched considerably, and the ore was lacking or of too low grade to be mined profitably at depths varying from about 2,000 to 3,000 feet. It would be folly to generalize from these facts, however, and to claim that profitable ore will always extend downward to a depth that is between one tenth and one seventh of the length of the outcrop. In fact the only generalization that can be offered on this topic is the unwelcome one that, in the great majority of cases, the depth to which profitable ore extends is very much less than the total length of a vein, and that the ratio between these two dimensions varies greatly in different deposits, and even in different parts of the same deposit.

The Presence of Certain Species of Plants not a Very Reliable Indication of the Existence of an Outcrop

The assertion is sometimes heard that it is possible to locate veins or other ore deposits by seeking for relatively closely crowded plants of certain species, and even that numerous examples of one species flourish over one kind of ore, while another species marks the outcrop of other minerals.

That there is some slight ground for these ideas is no doubt true. It has long been known, for instance, that *Viola calaminaria,* the "zinc violet," marks the outcrops of zinc deposits in Westphalia, and R. W. Raymond, in a paper read before the American Institute of Mining Engineers in 1886, reported that he had been told that the same plant had been recognized at the Horn Silver Mine in Utah, the ore of which contains considerable zinc sulphide. He describes in the same paper *Amorpha canescens* or the "lead plant" that occurs most plentifully over deposits of galena in limestone in Michigan, Wisconsin, Illinois, and southwesterly therefrom, and names other species that have been suspected to be indicative of the presence of outcrops.

S. B. J. Skertchley in *The Tin Mines of Watsonville* states that a plant called *Polycarpea sperostyle* marks the copper deposits of North Queensland.

In spite of the reputed facts just stated it is evident that each such idea is in all probability applicable, if at all, only to a more or less restricted area. In a given district certain ore and gangue minerals under existing atmospheric conditions may, by weathering, form soil peculiarly favorable for the growth of a species of plant indigenous to the region. It is, however, extremely unlikely that all four of these factors, (1) character of the ore minerals, (2) nature of the gangue, (3) atmospheric conditions, and (4) the same indigenous species of plant, will exist together in many other localities.

Some veins, usually at low elevations, are undeniably the loci of active water courses, and their outcrops may be marked by bands of unusually luxuriant vegetation. Such bands may or may not contain unusual proportions of one or more plant species. The same phenomenon may be found, however, whereever a deep, unmineralized fissure penetrates a porous stratum saturated with water under sufficient hydrostatic head to force it to the surface. It is also true that, if a vein at a relatively high elevation has been fractured by post-mineral movements, and left in a comparatively porous condition, it may drain rain water rapidly away from the outcrop, and thus produce a band of stunted or scanty vegetation. The outcrop of a dike of igneous or a stratum of other rock that on weathering yields a soil containing a deficiency of the substances required to support plant life may evidently, however, be marked in the same way.

Bands or patches of unusually luxuriant, stunted, or relatively scanty vegetation may, then, be worthy of a prospector's attention, but no great reliance should be placed on such phenomena. The idea that outcrops of certain kinds of ores are marked by peculiar species of plants is even less generally reliable, although doubtless applicable in more or less restricted areas.

Additional Observations on Outcrops

Zinc is the only metal that is commonly so completely leached from an outcrop as to leave no remainder there. More or less silver usually remains behind as horn silver. Gold disappears from the surface only when certain rather unusual conditions, outlined in an earlier chapter, prevail, and some of the metal usually remains in the oxidized material even then. Lead is rarely transported to any extent, its oxidized forms are rather easily recognized, and the relatively great weight of all lead minerals suggests the presence of that element. Some blue or green copper stain nearly always remains near the surface where that metal was originally present as primary ore. Other metals that leave stains in the oxidized portions of ore deposits are chromium (green), cobalt (lilac), manganese (black), molybdenum (yellow), and iron (yellow, brown, or red). It should not be expected, however, that the vestiges mentioned will always remain exposed directly to the air. It is usually necessary to sink from a fraction of an inch up to several feet below the surface before they appear.

The presence of iron oxides, of yellow, brown, or red color, in an outcrop has far more significance than merely to indicate that the primary ore contained some combination of iron and sulphur; they serve as evidence that metals other than iron have been leached out of the material. The truth of this statement can be appreciated when it is known that ferric sulphate is a most important solvent of metals in oxidizing ore bodies, and, when ferric sulphate reacts with a metallic sulphide, the metal in that sulphide is dissolved and some ferrous compound of iron is formed. Where the conditions are strongly oxidizing, as they are in an outcrop, such ferrous compounds are changed to insoluble ferric oxide in the presence of much copper. If a small quantity of copper is present, part of the ferrous iron will be converted to ferric oxide. If a strongly neutralizing substance, such as calcite (calcium carbonate), is present, a ferrous compound may be changed to iron hydroxide or iron carbonate, and oxidation of these substances may also produce insoluble ferric oxide. Moreover, the decomposition of several sulphides of metals may also result in the formation of some iron oxides. Iron oxides in an outcrop may, then, prove that the deposit once contained metals other than iron, although most or all of them have been leached out of the oxidized material.

As a result of studies made by Augustus Locke, Roland Blanchard, P. F. Boswell, and others, it is now often possible, by carefully scrutinizing the physical characteristics of the iron oxides in an outcrop, to determine just what sulphide minerals were originally present at the points where the iron oxides are now found. It has been learned, for instance, that the iron oxide deposited when chalcocite (sulphide of copper) goes into solution differs markedly from the oxide of iron that remains when chalcopyrite (sulphide of copper and iron) decomposes. A person familiar with the criteria involved can determine with a considerable degree of accuracy what sulphide minerals were originally present even when no other trace of a mineral originally present as a sulphide is left in an outcrop, as might be the case if the primary sulphide were sphalerite (sulphide of zinc).

The varieties of iron oxide that may be found in a leached ore body are numerous and not easily distinguished without protracted study so it is impracticable to include many data concerning them here, but there follow some criteria that relate to disseminated copper deposits, which have been abstracted from Augustus Locke's *Leached Outcrops as Guides to Copper Ore.*[5]

Disseminated copper deposits, before exposure to oxidation, contain specks, patches, and occasional veinlets of copper and copper-iron sulphides in a rock gangue. While the grade of the material is low the deposits are often of enormous size and can be worked very profitably when copper sells at a reasonable figure.

When such deposits are exposed to oxidation the copper may be almost completely leached out, only blue or green stains being left here and there in the material. It is then important to determine whether the sulphide material originally present contained considerable copper or was almost barren pyrite. If evidence can be found that a quantity of copper has been removed from the outcrop by leaching, it may pay to prospect material that is worthless where exposed near the surface, by drilling or in other ways.

It has already been stated that iron oxides are usually left in the cavities originally occupied by sulphides if the latter are attacked and dissolved. If cavities once occupied by sulphides are filled with iron oxides, considerable copper was originally present, and it is often possible to determine what the original mineral was by the physical characteristics of the iron oxides. In general it may be said that, if the areas, once sulphides, are completely filled with iron oxides, the original sulphides contained more copper than was present if cavities are filled with angular cells with thin walls or are merely lined with coatings of iron oxides. If, however, the cavities are empty, but are surrounded

---

[5] The Williams and Wilkins Co., Baltimore, Md., 1926.

by borders of iron oxide in contact with the cavities or separated therefrom by unstained areas, the primary sulphide was predominantly pyrite and the primary ore is probably worthless.

In general it may be said that the less movement of iron out of and away from the cavities there has been, the higher the percentage of the copper in the ore. It is also generally true that iron oxides that have reddish colors are more favorable to there having been a relatively high percentage of copper in the primary ore than is yellow or yellowish brown material.

The principal annunciated (no or very restricted movement of iron from the position where copper or copper-iron sulphide was originally deposited means that the ore once contained a relatively high proportion of copper) may be extended, to some extent, to vein deposits, and further investigation of this very interesting subject is providing us with other data which will aid us greatly in our estimation of the potentialities of a deposit of which only the outcrop can be scrutinized.

From what has been said it should be apparent that iron and other stains and porosity are favorable features in an outcrop that has been opened to a depth of several feet, and, without them, additional work is rarely justified unless gold alone is the metal sought and occurs in the oxidized vein material.

CHAPTER VII

THE BLOWOUT MYTH

Foreword

From time immemorial the speech of miners has contained frequent references to "blowouts," yet this is a term that is almost never used by geologists or mining engineers. All mining men have, however, heard tales of great quartz blowouts, iron blowouts, etc., and the term has become so firmly established in the vocabulary of prospectors that it will probably never be displaced.

A Blowout Defined

One seeks in vain through economic geologies for a definition of this word. In fact it is rarely seen in print. One of the very few available definitions is contained in Bulletin 95 of the U. S. Bureau of Mines, by Albert H. Fay, entitled "A Glossary of the Mining and Mineral Industry." In this bulletin occurs the following sentence: "A large outcrop beneath which the vein is smaller is called a blowout."

While the statement just given is probably true it does not tell the whole story. Usually the term blowout is applied to any relatively large outcrop of ore or gangue minerals or both, where the length is not very much greater than the width. Sometimes similar masses of minerals that are not commonly associated with ore as gangue are also thus designated by prospectors who fail to recognize the barren nature of the material. Most so-called blowouts outcrop prominently and attract attention for this reason, as well as because of their unusual width.

The idea that seems to be ordinarily held is that volcanic forces have pushed or blown out these blowouts, presumably in a molten condition, from the deep interior of the earth through a more or less cylindrical or tubular passageway. This is certainly a fallacious notion in most instances, and indicates a lack of understanding of the principles covering the deposition of the commoner ore and gangue minerals.

Few Mineral Deposits Were Originally Molten

With the exception of one or two unusual classes mentioned later, no deposits of ore or gangue minerals were extruded in a molten condition. On pages 7-9 will be found a brief outline of the manner in which it is confidentially believed most ore deposits were formed. The possibility that metallic gold and other valuable ore minerals may have arisen in a melted condition and been deposited by cooling is not even mentioned there and such a general theory of the origin of ore deposits must be discarded as utterly untenable. Only the peculiar types of deposits mentioned later in this chapter have had any connection with volcanic activity. Other types were formed by deposition from water solutions, beneath the surface, as has already been stated, and have not been "blown out" in any way.

"Burnt" Outcrops Not Necessarily Indicative of Volcanic Activity

"But," some readers may say, "how can you explain the brick-red, yellow, or brown, porous, cindery appearance of many outcrops of blowouts and other mineral deposits if volcanic activity has not entered into their formation?" In order to understand this matter it should be remembered that combustion is merely a process of oxidation. When wood, coal, etc. are burned, atmospheric oxygen combines with certain constituents of the fuel, and gaseous products (carbon dioxide, carbon monoxide, and other substances) pass into the air, and the chemical reactions involved generate heat. If there is any unconsumed residue (ashes or cinders), it will usually be porous, and will be stained by iron oxides if the fuel contained iron and if combustion has been fairly complete. Any outcrop that contains oxidizable minerals is susceptible to a slow form of combustion called oxidation. Atmospheric oxygen attacks and combines with the iron that is apparently locked up so securely in the silicates and sulphides, and converts it to ferric oxides of red, yellow, or brown tints. Atmospheric carbon dioxide and water attack other ingredients and form soluble carbonates or soft, easily disintegrated hydrous silicates. Such carbonates and hydrous silicates may be dissolved or removed mechanically, and the residue then has a highly iron-stained, porous appearance quite similar to that of cinders or the surface of a lava flow. To be sure, the process outlined frequently takes place with extreme slowness, and the heat generated is usually so rapidly absorbed by the surrounding rocks or so quickly radiated into the atmosphere that its development escapes attention. Likewise any gases generated are immediately dissipated and are unnoticed. Heat and gases are generated, however, just as when fuel is burned, and the whole process of oxidation by weathering is quite analogous to ordinary combustion. As a matter of fact gases, at least, are in some instances generated so rapidly through weathering as to be quite easily noticed. Not infrequently the odor of burning sulphur is apparent around dumps containing a high proportion of pyrite (iron sulphide). In some places, indeed, the sulphur in pyrite is largely eliminated and the ore converted to oxides by "heap roasting," which is simply a natural process in which weathering is utilized to convert sulphides to a more valuable, oxidized product that possesses many points of similarity to material that has been roasted artificially. Some wood or other fuel is used, however, to start or speed up the reactions.

The important point to be noted in the preceding paragraphs is that the products of weathering and combustion are sufficiently alike to explain the erroneous idea that heat has been a direct factor in the formation of many blowouts and other types of ore deposits.

Ore Deposits Not Invariably Precipitated from Water Solutions

It is true that some ore deposits have been formed directly by the solidification of masses of molten earth material, but such deposits are relatively rare, and, with a few notable exceptions, unimportant. They are termed magmatic deposits. With the possible exception of some quartz veins they do not include the commoner types of deposits such as veins, shear zones, contact deposits, metasomatic replacements, bedded deposits, etc.

While it is true that practically all igneous rocks (solidified magmas) contain variable proportions of precious and base metals, the former are normally represented by mere traces, and the latter are usually present in such forms that their profitable extraction is impossible. Sometimes, however, certain valuable minerals, especially compounds of base metals, segregate or become concentrated at various points in the igneous mass instead of remaining uniformly distributed throughout the magma. This result is brought about by a process called differentiation, and under favorable conditions it may cause the formation of valuable accumulations of ore minerals. Such deposits are often of roughly lenticular form, but they may be quite irregular in shape and are more or less completely surrounded by igneous rock into which they usually grade insensibly. The gangue minerals are those that constitute the igneous rock in which a magmatic deposit is found. Deposits of this type are frequently confined to certain positions in an igneous mass, as near its borders, but they may also occur in the interior.

As an illustration of magmatic deposits may be mentioned the many occurrences of chromite (chromium-iron oxide) in peridotite or its alteration product, serpentine. When chromite is found in serpentine it may have been concentrated along with magnetite (magnetic iron oxide) during the alteration process by which the change from peridotite to serpentine took place, but many chromite deposits are confidently believed to be of purely magmatic origin. Deposits of ilmenite (titanium-iron oxide) and titaniferous iron ores in gabbro or similar rocks; of magnetite in granitic, syenitic, or dioritic rocks; of pyrrhotite (sulphide of iron), often containing nickel, in gabbroid rocks; and of chalcopyrite (copper-iron sulphide) and other sulphides in several types of igneous rocks have certainly been formed in this way, as have the deposits of platinum with chromite and magnetite in basic igneous rocks in the Transvaal, South Africa.

A few deposits of nonmetallic minerals, particularly diamond in peridotite and corundum (sapphire, ruby, etc.) in rocks ranging from syenitic to gabbroid, also have a magmatic origin.

It has been claimed by some geologists that the deposits of gold in quartz and alaskite at Silver Peak, Nevada, were produced by magmatic differentiation, but there seems to be some doubt about the correctness of this view. At any rate it is certain that almost no important deposits of gold, silver, lead, and zinc, and very few of copper, are of magmatic origin.

Have all valuable deposits of minerals that are primary ingredients of igneous rocks been concentrated by differentiation? No, some igneous masses, especially dikes, contain a sufficiently high proportion of disseminated ore minerals to be workable at a profit. The rocks involved are usually pegmatites, however. The ore minerals found in dikes of such rocks comprise cassiterite (oxide of tin); wolframite (tungstate of manganese and iron) and other tungsten minerals; columbite (columbate of manganese and iron); tantalite (niobate of manganese and iron); several yttrium, thorium, and cerium minerals; spodumene (silicate of lithium and aluminum) and other lithium minerals; molybdenite (sulphide of molybdenum); etc.

Gem stones, such as aquamarine and other varieties of beryl, quartz, topaz, tourmaline, etc. are also frequently found in pegmatites, as is also white mica or isinglass.

It should probably be mentioned that some pegmatites are transitional into quartz veins composed usually of that variety of the mineral known as "bull" quartz. It is somewhat difficult to decide whether quartz in such veins has been formed by the cooling of molten material or has been deposited from water solutions. In fact Pirsson says (*Rocks and Rock Minerals*, page 179): "It may be regarded as almost certain that no sharp line can be drawn between igneous fusions of silicates (molten silicate magmas) containing water under pressure and hot water solutions. It appears that under pressure water will mix in all proportions with magma so that at one end are molten fusions, and at the other, hot solutions."

In spite of the rather long list already given of substances that occasionally occur in economically important quantities in rocks that were once molten, such deposits are relatively scarce, and, as has been mentioned, very rarely contain valuable deposits of the metals sought by most miners. Igneous rocks that do carry unusually high proportions of metallic ores or valuable nonmetallic minerals might possibly be termed "blowouts," but it is a fact that this term is almost never applied to such masses.

The Facts About So-called Blowouts

There is no doubt but that most so-called blowouts contain ore and gangue that were deposited from water solutions in the same way that most of the commoner types of ore deposits were formed. The process was doubtless usually a gradual one; the temperatures of the solutions were below the critical temperatures of water under the pressures involved, and may in many instances have been little if any above surface water temperatures. There was nothing violent about the process, which would suggest or justify the use of the term blowout for the resulting deposits.

It is a fact, however, that those who have not given much study to the science of geology are prone erroneously to ascribe a violent origin to almost all natural phenomena. Mountains are supposed to have been thrown upward bodily in a very brief space of time; great canyons are thought to have been rent asunder by mighty and mysterious forces beyond the understanding of man; continents are believed to have been elevated or depressed; and blocks of the earth's surface are assumed to have been faulted upward or downward with great suddenness. Extreme ideas like these that ascribe all geological phenomena to cataclysmic forces are undoubtedly fallacious. The same very slow elevation and depression of land masses, the same very gradual fault movements, and the same protracted carving out of mountains, valleys, and canyons by running water and the wind that are occurring now have developed most of the land forms and rock structures that many people assume are caused by sudden and violent cataclysms. It is no doubt true that, at some periods in the past, the forces active in modifying rock structures and molding the earth's surface have been accelerated and at other times retarded, but in general it seems certain that most geological phenomena have resulted from the action of forces extending over enormous stretches of time. This statement may certainly be applied to most ore deposits; there is nothing about their origin that justifies their designation by terms so suggestive of violent and quickly consummated activities as does the word blowout.

The roughly circular or short and wide outcrops of ore shoots, ore bodies, or gangue minerals commonly called blowouts may have their positions fixed in so many ways that it is difficult to mention all of them; but most such masses are attributable to one of the causes listed below:

1. The intersection of two veins.
2. The intersection of a vein and shear zone.
3. The intersection of two shear zones or two faults.
4. The filling by deposition of a pipe-like channel produced by faulting along an irregular fissure.
5. The intersection of a relatively thin, highly inclined stratum of easily replaceable rock by a vein or shear zone.
6. The outcropping of a "chamber deposit" (cave filling) in limestone.
7. The outcropping of a "pocket," "lens," or other irregular mass of ore such as is frequently formed in limestone by metasomatic replacement near the contact with an igneous rock.
8. The filling of a volcanic vent by deposition from solution.

As the first five of the suggested possible causes for the development of blowouts also fix the position of some ore shoots, which will be discussed in a subsequent chapter, no further reference will be made to them at this time. The remaining three suggested causes are discussed briefly below.

Chamber deposits are relatively rare, and probably only a small proportion of the few outcrops of such deposits have peculiarities that would cause a prospector to call them blowouts. Such deposits are usually composed very largely of relatively soft and easily alterable minerals, so they rarely outcrop prominently. For these reasons, few, if any, so-called blowouts are of this type.

The development of contact deposits by metasomatic replacement of limestone intruded by igneous rock is a common cause for the development of many valuable ore deposits, especially of copper. Iron is usually associated with the copper in the form of pyrite (sulphide of iron), chalcopyrite (copper-iron sulphide), bornite (copper-iron sulphide), and hematite (oxide of iron), and the weathering of such ore leaves a residual mass of more or less impure iron oxide on the surface. Such so-called "iron hats" have not infrequently been termed "iron blowouts."

The metamorphosis of impure limestone by the heat and mineralizers expelled from intrusion of igneous rocks frequently results in the formation of masses of garnet, epidote, or other hard silicates that are resistant to weathering. Outcrops of such material are also often called blowouts.

While no volcanic neck or vent is positively known to have been mineralized by deposition from solution, the valuable ore in the Bassick Mine near Rosita, 7 miles east of Silver Cliff, Colorado, is believed by many geologists to have originated in this way. There the ore occurred coating boulder-like fragments of rock that partially filled a nearly vertical "chimney" with an oval cross section, which varied from 20 to 100 feet in diameter, and which was mined to a depth of over 800 feet. The material between the ore-coated rock fragments was a soft alteration product of some light, porous, volcanic substance that evidently was readily penetrated by the solutions from which the ore was deposited. The deposit in the Sliver King Mine, north of Superior, Arizona, is in the form of an irregularly cylindrical chimney, pipe, or neck, and many miners would have called its outcrop a "blowout."

With reference to the definition of a blowout given by Mr. Fay, already quoted, it may be said that the lack of pressure near and at the earth's surface undoubtedly makes it easy for much rock fracturing to occur there, even when the forces involved cause the formation of a single fissure or narrow shear zone at greater depth. Subsequent mineralization of the feeding channel and of the enlarged zone of fractured material developed at or near the surface will cause the formation of a blowout as defined by Mr. Fay.

From what has been said it should be quite evident that few or no so-called blowouts have been blown out or expelled violently in any way. They are merely deposits of ore or gangue of ordinary type and origin although possessed of a somewhat unusual shape. The term is then a poor one, is quite superfluous, and should be dropped from the vocabularies of all who desire to seem well informed concerning ore deposits.

CHAPTER VIII

SOME FALLACIES RELATING TO ORE SHOOTS

An Ore Shoot Defined

An ore shoot is a body of relatively high-grade ore in a deposit of low-grade ore or barren minerals. If small, an ore shoot is often called a "pocket," "bunch," or "lens," and, less commonly, a "nest"; larger shoots are frequently termed "pay streaks," or, more rarely "ore courses." While a mass of ore surrounded by barren country rock is considered to be an ore shoot by most writers and miners, such a chimney, pipe, or neck has been discussed elsewhere in this book, and it seems best to exclude it from the present discussion. Other masses also sometimes called ore shoots, which will be excluded from consideration in this chapter as well, are the relatively wide or thick parts of a deposit no richer than the adjacent narrow or thinner portions, and all bodies of rich ore formed by secondary enrichment. The term ore shoot will, then, be used in a very restricted sense in this chapter, so as to include only masses of relatively high-grade primary ore that occur within bodies of lower-grade ore and gangue minerals.

The primary ore in most types of ore deposits varies in grade from point to point and that shoots of rich ore occur in leaner material are facts familiar to practically all miners and prospectors, as was stated in the first chapter. If this condition did not exist—if ore deposits were of uniform grade throughout, a very large proportion of all known ore deposits would long since have been exhausted, there would be little work for mining geologists to do, and mining would lose much of its glamour and mysterious fascination. The fact that most valuable primary ore is concentrated into shoots tends to make mining more or less of a gamble, and a majority of men are gamblers at heart. The absence of ore shoots would certainly take most of the risk out of mining and reduce the amount of money lost in the industry to a comparatively low figure. In fact mining and manufacturing would then be on about the same plane; dishonesty, incompetency, and the operation of the law of supply and demand would be the only factors that would cause failure.

Since most of the profit made in mining primary ore has been derived from ore shoots, it is highly important that all miners should be familiar with their nature and the conditions that have brought them into being.

The Nature of Ore Shoots

Penrose[6] says: "The typical ore shoot is of a more or less columnar shape, dipping vertically or at a steep angle; but most ore shoots are less regular in form, and some are strikingly irregular, protruding and receding on all sides, thining down in some places to a narrow neck, in others widening into great dimensions, and throwing out long arms from the main body. Some ore shoots, instead of dipping steeply, lie almost horizon-'ally or meander up and down. . . . Some shoots form lenticular masses, following a fissure for many feet along its course, and finally thinning out at either end into leaner ore or gangue. . . . Ore shoots may crop out at the surface or may be found at a greater or less distance below. In depth, they may continue for great distances, or they may terminate at less depth, and may or may not be succeeded by other shoots below."

Ore shoots in general may occur singly, or there may be several in the same deposit. In the latter case they may be at the same or at different depths, close together or far apart, parallel or nonparallel. The ore may be of rather uniform grade throughout the shoot, or the value may vary from point to point.

Shoots may extend downward a few score or 2,000 or more feet, and their other dimensions are equally variable. Vertical or steeply pitching shoots always terminate downward at some point, but smaller shoots are likely to be found below the end of a large one.

Ore shoots in a vein or some other type of tabular deposit may pitch or dip downward parallel to the dip of the deposit, but in most instances they pitch more or less steeply to the right or left of an observer facing in the direction toward which the deposit dips. Such shoots may be richest or poorest where widest. It is often stated that shoots are apt to pitch in the direction of any grooves or scratches found on the "slickensided" enclosing walls if deposited in a fault fissure, but they certainly do not always do so.

One might gather from what has been said that the position of an ore shoot is a mere matter of accident, but such is not the case. They are formed through the operation of perfectly definite natural laws, and several possible causes of their formation are given below.

Conditions Favoring the Formation of Ore Shoots

1. Intersections of Veins, etc.—The intersection of two veins, a vein and a shear zone, a vein and a fault, a vein and a fahlband, or even a vein and a zone of unusually numerous cracks or joints (cross-fractures), as well as some other combinations of the phenomena mentioned, is not infrequently marked by the existence of a valuable ore shoot produced by the mingling of solutions, the crushing and consequent easy impregnation or re-

---

[6] R. A. F. Penrose, Jr., *Types of Ore Deposits*, p. 325, 1911.

placement of the wall rock, the partial obstruction to the passage of solutions, or by other causes. Such a shoot seems especially apt to form if the angle of intersection is acute. Good ore shoots are so frequently found at such intersections that they are usually considered to be worth prospecting even though little or no evidence of the existence of such shoots is found on the surface of the ground.

2. Presence of Certain Wall Rocks.—If a vein or some similar tabular deposit intersects an easily replaced rock, such as limestone, or one that has a strong tendency to cause the precipitation of certain substances brought into contact with it in solution, such as shale or slate containing organic material, an ore shoot may form at such an intersection. The same result may be brought about where a vein or similar tabular deposit passes through a mass of relatively brittle, easily fractured rock, such as quartzite, if elsewhere it lies in tougher material, such as mica schist.

The two possible causes already mentioned are probably responsible for the existence of the majority of all ore shoots.

3. Crushed Areas Not Produced by Intersections.—If the country rock adjacent to a certain portion of a deposit is unusually crushed, an ore shoot may develop there. This case differs from the preceding one in that there is no change in the nature of the wall rock when passing from the unshattered to the crushed area. It is sometimes difficult to understand why the wall rock should be badly broken at one point and unbroken elsewhere.

4. Closely Approaching Veins.—If two veins curve toward each other and are close together for a relatively short distance, the rock between them where they approach most closely may be greatly fractured, and this may lead to the development of an ore shoot at that point.

5. Swells and Pinches Produced by Faulting.—If a crack or fissure is not perfectly straight and one side is moved or faulted relative to the other, the result is often a series of unusually wide openings (swells) separated by narrow stringers (pinches). Ore shoots may develop in either the swells or the pinches.

Many other possible causes of the development of ore shoots might be given, but it would serve no good purpose to do so. As a matter of fact, only the first, second, and fourth of the causes already briefly discussed bring about conditions that can easily be used by a prospector seeking to locate ore shoots on the surface.

Although five different conditions sometimes favoring the formation of ore shoots have been mentioned, it should not be supposed that such shoots are usually produced by a single distinct cause. Doubtless more than one, perhaps several, causes have combined to form an ore shoot in most instances, and, as Penrose[7] says: "The evidence of some of these causes may have been

---

[7] *Op. cit.,* p. 326.

much obscured or even obliterated since that time, so that the determination of just what cause has been uppermost in influence is often impossible. Thus a given shoot may clearly owe its presence, in part at least, to the replaceable character of the wall rock, but, if the structural features in the same spot were also favorable to ore deposition, it might be difficult to decide which influence had been the more important."

Following this brief statement of facts about ore shoots, it is desirable to consider some of the common fallacies relative thereto.

Veins and Other Mineralized Structures do not Always Contain Ore Shoots

A fallacious idea that has cost prospectors very dearly is the notion that every vein, shear zone, fahlband, etc. contains at least one shoot of good ore. In the minds of most miners this idea amounts to a positive conviction, but it is absolutely erroneous and abandoned shafts and tunnels scattered over a thousand hills are monuments to this fallacy and to the misguided efforts of those who dug them. Did they stop too soon? Perhaps so in some instances, but it is an indubitable fact that typical fissure veins filled with gangue minerals commonly associated with good ore have been found to be totally lacking in ore shoots after extensive development. In some instances they have produced sufficient barite, fluor spar, pyrite, or some other commercially important substance to justify opening them extensively, so it cannot justly be claimed that the search for the ore shoot was not carried far enough. Perhaps the peculiar conditions that cause the formation of ore shoots were not present when such deposits originated; possibly the solutions from which the gangue or relatively valueless metallic minerals were deposited contained no valuable ore; but, no matter what the explanation may be, it is certainly true that such shoots are the exception rather than the rule. The possibility of finding a rich ore shoot is the lure that tempts many a poor miner to drive his tunnel in gangue and ore of too low grade to be mined profitably. His faith, optimism, and perseverance may be admired, but his judgment must be questioned. It is undeniably true that he has sometimes succeeded even when no one would have been justified in predicting his success, and such successes are widely advertised, but the hundreds of failures endured by his equally persevering, but less lucky, fellows are rarely mentioned and soon forgotten. On this subject Wallace[8] says: "Lodes which, after proper surfacing, do not reveal reasonably fair values in at least one place, are seldom worth the holding by a poor man. Blind shoots may be discovered by sinking at random, but this is uncertain. Depth may show better values, but generally it does not. Even if this were known to be true in a particular case, how could a lean purse

---

[8] J. P. Wallace, *Ore Deposits for the Practical Miner*, Hill Publishing Co., pp. 207-8, 1908.

drive the drill to the required depth? Mining outside of pay-shoots had better be left to him who hath and to spare."

Ore Shoots Not Always Found Where Conditions Seem Favorable

It should not be supposed that ore shoots have always been formed even where the conditions appear to have been most favorable for their development. One vein may intersect another without appreciable influence upon the grade of the ore at such intersection, and a vein often cuts through limestone or carbonaceous shale without apparent effect upon the grade of ore deposited adjacent to or in such rocks. As a matter of fact the factors mentioned as favorable to the formation of ore shoots fail to produce richer ore much more frequently than they cause the deposition and concentration of such ore. This is equivalent to saying that the great majority of vein intersections, masses of brecciated country rock intersected by veins, etc. do not mark the position of ore shoots. It is furthermore true that conditions listed as among those that often lead to the development of ore shoots not infrequently seem to have a totally opposite effect; a vein, for instance, may show a decrease of values at the point where cut by a second vein. In spite of what has just been said, it is still true that the conditions mentioned as likely to cause ore shoots are much more apt to have this effect than to bring about a lowering of the values.

Most mining operations are more or less of a gamble even when the conditions appear most favorable, and it is usually considered a good policy to prospect thoroughly such portions of a mineral deposit as experience has shown are most apt to contain ore shoots. To drive or sink blindly in the hope of encountering such a shoot is an extremely hazardous proceeding, but to reach two vein intersections or some other favorable geologic condition may be regarded as a commendable enterprise, even though it must be admitted that failure will result more often than success. On this point Penrose[9] says: "Though the occurrence of an ore shoot, even under conditions most favorable for its formation, is the exception and not the rule, yet it is worth while to study the causes that produce it, and to look for ore under similar conditions elsewhere. The search in many cases will be fruitless, but another 'exception' exists somewhere, and a knowledge of the conditions under which it may occur is helpful. Just as the pearl in the oyster is an abnormal segregation resulting in a beautiful gem, so the ore shoot in the earth is an abnormal segregation resulting in precious minerals; just as there are thousands of oysters that contain no pearls to one that does, so there are thousands of apparently favorable receptacles for ore shoots that yet carry none; just as the pearl diver finds it remunerative to hunt for the oyster that may contain his prize, so the miner finds it remunerative to hunt for the spot that may

---

[9] *Op. cit.*, p. 354.

contain his ore; and though the paths of both are strewn with disappointed hopes, yet the possibility of realization leads them on."

Penrose was apparently unaware of the fact that valuable pearls are never found in oysters, but only in other mollusks, yet there is no doubt as to the idea he intended to convey.

Conditions Favoring the Development of Ore Shoots in One Locality May Not Do So Elsewhere

In other chapters mention has been made of the unfortunate fact that prospectors almost invariably assume that, because a given set of conditions accompany good ore in one place, the same conditions found elsewhere can be accepted as an almost certain indication of the presence of such ore there. Such assumptions are especially dangerous because they are half-truths. If it were impossible to reason from cause (geologic conditions) to effect (the presence of ore), geology would be of no service to miners, and the tenderfoot would be just as likely as the experienced miner to locate an ore shoot. There are such things as favorable and unfavorable indications to guide the seeker for ore, and as geologic science progresses such indications will be better understood and the number increased. It is probable that many very important factors influencing the deposition of ore are, however, of such a nature as easily to escape the notice of anyone not capable of using the compound microscope, chemical apparatus, and other tools unfamiliar to prospectors. The miner is too apt to generalize and he often utterly fails to appreciate the many factors that may in one locality modify or nullify the apparent result of the conditions he has observed elsewhere.

Certain it is that no miner is justified in claiming that conditions favoring the formation of ore shoots in one locality will certainly have the same effect many miles away. They may do so, it is true, but no one should express great faith in the beneficial effect of such conditions until the matter has been investigated by actual mining operations.

It is undeniable that in a restricted area the geologic phenomena that caused the formation of an ore shoot at one point are very apt to produce a similar result if repeated elsewhere; so a knowledge of such favorable factors is often very useful. As the search for ore is extended farther and farther from the restricted area, however, the criteria applicable there become less and less reliable, and, when the prospector has traveled some distance from the small area that he has studied, he should not place much reliance upon the conditions known usually to accompany ore shoots there. In the absence of any other guide in the search for ore, he will be justified in investigating points where such conditions are well developed, but he will probably suffer great disappointment if he places too much faith in them.

No Fixed Relation Exists Between the Horizontal and Vertical Dimensions of Ore Shoots

The ratio between the greatest horizontal and vertical dimensions of an ore shoot is by no means a fixed quantity. In flatly pitching shoots or in shoots in blanket deposits, the greatest horizontal length may be many times the greatest vertical thickness. Steeply pitching shoots in veins or other tabular deposits usually extend downward a distance at least twice as great as the greatest horizontal dimension. If, however, an ore shoot outcrops on the surface, much of it may have been removed by erosion, and in this case it is unsafe to assume any considerable downward extension. According to Lindgren,[10] H. C. Hoover, from an examination of seventy mines, concluded that ore shoots are generally lenticular, and that a given shoot can be assumed to extend below any level a distance at least equal to·one half of its greatest horizontal dimension at that level. Lindgren follows this statement by saying that at Cripple Creek he and Ransome found that the shoots that began distinctly below the surface (blind shoots) had a downward length varying from one and one half to five times that of their greatest horizontal dimensions. He further states that primary shoots rarely continue for more than 2,000 feet along the strike, or extend downward along their pitch for more than the same distance.

Shoots in veins very rarely extend downward a distance equal to ten times their greatest horizontal dimensions. In fact it is likely that the majority are less than five times as long down their pitch as they are horizontally. Experience has shown that no one is justified in assuming or claiming that the pitch length of a blind shoot will be more than three or four times its horizontal length, although it is by no means impossible that the ratio will be found to be greater than this. Where the upper part of a shoot has been partially removed by erosion, still more conservative estimates must be made, as previously stated.

The figures given as applicable to Cripple Creek show that the ratios between the greatest horizontal and the pitch lengths of shoots are likely to be quite different even in a relatively small area. Perhaps in some districts these figures are more nearly equal, but it would be folly to assume such equality until the extensive development of a number of properties has shown it to exist.

---

[10] W. Lindgren, *Mineral Deposits*, pp. 186-7, 1919.

CHAPTER IX

THE ALUMINUM MYTH

An Erroneous Idea

Many miners believe that, if a mineral or rock contains a considerable percentage of a certain element, it is necessarily an ore of that element and is therefore valuable. Nothing could be further from the truth. Unless the element under consideration can be extracted so cheaply that the selling price will yield a profit after all costs have been deducted, the material is not an ore of the element under consideration.

The idea that all minerals that contain a considerable proportion of a valuable element are ores of that element has been called the "aluminum myth" by the writer for the reason that the element involved is so often aluminum. Frequently, samples which the senders have learned contain aluminum are submitted to the Arizona Bureau of Mines with requests for information as to where the material can be sold, and a reply to the effect that it is worthless always arouses feelings of surprise and disappointment.

The Facts About Aluminum

Aluminum is the commonest metal in the earth's crust. In fact it has been estimated that the percentage of aluminum in the solid crust is 8.16 while that crust contains only 4.6 per cent of iron.[11] All clays, shales, most igneous rocks, and many soils contain considerable percentages of aluminum and smaller proportions of this element occur in other rocks. In the report, "The Clays of Eastern Colorado," made by the writer and published by the Colorado Geological Survey in 1914 as Bulletin 8, are given analyses of 115 clays and shales. They contain from 7.6 per cent to 35.7 per cent of aluminum oxide (Al_2O_3) which is known as alumina. The average alumina content of all samples analyzed is 17.86 per cent. Since alumina contains 53 per cent of the metal aluminum, there is 9.46 per cent of this metal in a clay that contains 17.8 per cent of alumina, while a 35 per cent alumina shale will contain nearly twice as much aluminum.

Excluding peridotite and other relatively rare, very basic species, igneous rocks contain, on the average, almost exactly the same percentage of alumina as do clays and shales, and even garden dirt is apt to carry a considerable percentage of alumina.

---

[11] F. W. Clarke, *Analyses of Rocks*, Bul. 163, U.S.G.S., p. 15. 1900.

In spite of the widespread distribution of alumina, very few minerals have been or are being used commercially as sources of this metal. The mineral first to be so used is cryolite, a sodium-aluminum fluoride ($3NaF.AlF_3$) which contains only 12.8 per cent of aluminum when pure, but from which the metal may be easily extracted. The only known economically important deposit of cryolite is in Greenland, however, and this mineral is now only a minor source of aluminum although still used as a flux in the treatment of other aluminum ore.

Cryolite often resembles a piece of white ice, is easily scratched with a knife, and has some tendency to break in three directions which are nearly at right angles to each other.

With the development of cheap hydroelectric energy, bauxite became and remains the principal source of aluminum. Pure bauxite is a hydrous oxide of aluminum with the formula $Al_2O_3.2H_2O$, but commercial bauxite as mined varies in composition from $Al_2O_3.H_2O$ (diaspore) to $Al_2O_3.3H_2O$ (gibbsite) and it usually contains some ferric oxide, titanium oxide, silica, etc. as impurities, so the Arkansas bauxite contains, as an average, only about 54 per cent of alumina. Besides Arkansas, the mineral is being mined in North America in Georgia, Alabama, and Tennessee.

Bauxite is an earthy substance of white, yellowish, brownish, or reddish color, which smells like clay when moistened and is not always identifiable without laboratory tests, although the presence of what appear to be spheroidal pebbles of different color and hardness than the matrix is a rather characteristic feature. There is no market for bauxite since the producers of aluminum own their own deposits, but a large deposit of the mineral far from present sources, where cheap electric energy is available, might be valuable.

Good cryolite could probably be sold profitably, but the mineral is so rare that there is an exceedingly small chance that a deposit might be located by a prospector.

From what has been said it should be evident that cryolite was used as a source of aluminum because of the relative ease with which the metal may be extracted from it and that bauxite is now the principal ore of the metal because of the high percentage of aluminum that it contains. It is very unlikely that any other mineral that may be found in considerable quantities will have any immediate value as a source of aluminum, although efforts have for years been made to extract the metal commercially from clays and other substances that contain it. There is no doubt that aluminum can be obtained from many substances, but the cost of the operation is more than the product is worth. Probably a cheap method of extracting the metal from one or more of the common substances that contain it will some day be perfected, but when this aim has been accomplished the ore will have no value because of its commonness and the price of aluminum will be fixed solely by the cost of extracting it.

Other Valuable Substances Unrecoverable from Some of their Compounds

Iron.—Iron is an element of such small value that prospectors rarely assume that it is worth anything if analyses of minerals reveal its presence. It may be said, however, that there are scores of valueless minerals that contain more or less iron, and the only commercially important ores of the metal are the oxides and the carbonate which when pure contain 48 per cent or more of iron.

Magnesium.—This metal also occurs in a large number of valueless minerals. Its only commercial sources are, however, the carbonate (magnesite) and several soluble salts that contain chloride of magnesium.

Potassium.—Although potassium is an ingredient of orthoclase and microcline feldspars, which are plentiful constituents of granite, syenite, some porphyries, rhyolite, trachyte, and some other igneous rocks, as well as some gneisses, conglomerates, sandstones, clays, and other rocks, the metal cannot be extracted profitably from feldspars.

The principal commercial sources of potassium are several soluble salts in which it occurs as the chloride, but considerable of the metal is obtained from alunite (hydrous sulphate of potassium and aluminum) and there has been a little production from leucite (silicate of potassium and aluminum).

The chlorides are soluble and will probably not be found by prospectors on the earth's surface, so are not described.

Alunite is a white to pinkish mineral which can easily be scratched by a knife. Its luster is usually dull, but it is glassy when crystalline. It ordinarily occurs in rather porous, granular masses, but it is sometimes earthy. It is usually necessary to test in a laboratory material suspected to be alunite before it can be identified with certainty. Alunite also contains 37 per cent of alumina and it was recently announced that a process for extracting aluminum profitably from it has been perfected, but none of that metal is as yet being so produced.

Leucite is a colorless, white, or grayish mineral which is too hard to be scratched with a knife. Its luster is glassy and it occurs in spheroidal crystals (usually the unmodified isometric trapezohedron or tetragonal trisoctahedron) which often show octagonal cross sections in certain igneous rocks.

Unless potassium occurs in one of the forms mentioned, it cannot be extracted profitably, and its presence in other substances may be ignored.

Silica.—Silica (SiO_2) is oxide of silicon, and, as quartz, it has many uses, but it cannot be extracted profitably from the so-called silicates, nor can quartz be separated profitably from other minerals with which it is associated. Pure or nearly pure silica must occur unmixed and uncombined with other substances, as quartz or opal, to have any value, and the fact that SiO_2 is reported in the analysis of an ore may be ignored, except as it may affect the cost of smelting the ore.

Sodium.—According to Clarke[12] sodium is one of the eight elements present in the earth's crust in amounts exceeding 1 per cent, that crust containing 2.75 per cent of sodium. It is evident, therefore, that sodium must be a constituent of many minerals and rocks. The only commercial sources of the metal are, however, halite or rock salt (chloride of sodium) and three different carbonates of sodium which are found in the brines of soda lakes or in the deposits formed when such lakes have evaporated to dryness. If, then, soda (Na_2O) is reported in an ore analysis, it should not be assumed that its presence adds to the value of the ore.

Sulphur.—All of the sulphur of commerce is obtained from deposits of pure, elemental sulphur, and the sulphur that occurs in all sulphide ores has no value and its presence in an analysis of an ore may be ignored except as it may affect the cost of smelting the ore. It is lost in the smelter fumes unless plants for the manufacture of sulphuric acid have been installed in connection with smelters. An exception to the above statements is the mineral pyrite which contains 53 per cent of sulphur. It is mined to some extent as a source of sulphur that is used in making sulphuric acid.

The Myth Illustrated

A miner recently wrote the Arizona Bureau of Mines to the effect that he had a vein of ore that concentrated easily, and that he had crushed and panned enough material to have an analysis of the concentrates made. He said that these concentrates proved to have the following composition:

Copper	4.2	per cent
Gold	0.05	ounce per ton
Silver	15.2	ounces per ton
Zinc	12.3	per cent
Iron	28.5	per cent
Sulphur	30.1	per cent
Manganese	2.8	per cent
Arsenic and antimony	6.1	per cent

He further stated that he had submitted this analysis to a smelter and had learned that it would give him only about $8.00 a ton for such ore although he figured that the elements present in it, excluding the arsenic and antimony of which he did not know the proportion, were worth over $50 at current quotations. He denounced the smelter as a robber and advocated the establishment of a state smelter.

Of course, he was told that it was necessary to penalize him for the zinc, arsenic, and antimony, and that, although some credit was allowed for the iron and manganese, it did not equal

---

[12] *Op. cit.*

the value of these substances in elemental form, and that nothing could be paid for sulphur. He was assured that all elements could not be extracted profitably from such ore by anyone and that the offer of the smelter was a reasonable one.

Another prospector recently sent in some samples which he said had been analyzed with the following results:

SiO_2	58.21 per cent
Al_2O_3	11.23 per cent
Fe_2O_3	9.12 per cent
CaO	4.02 per cent
MgO	3.05 per cent
K_2O	2.33 per cent
Na_2O	1.12 per cent
Loss on ignition	8.20 per cent
Moisture	2.20 per cent
Total	99.48 per cent

He stated that he considered it evident that he had found some aluminum ore, and he wanted to know where he could sell it and be also paid for the mangnesium, potassium, and sodium in it which, he stated he had learned, were worth, respectively, about 32 cents, $19.50, and $4.25 a pound.

It was, of course, necessary to tell him that his material was worthless and that not a single element could be extracted profitably from it. Furthermore, the material, which was a sandy shale, was not even adapted to the manufacture of cheap red bricks, since when burned in an assay muffle it fused very suddenly when the color of the muffle was a deep orange, and, at a temperature very little lower than that of fusion, it had not burned hard enough to be handled without injury, so there would probably be much loss from over- and underburning.

These two illustrations should suffice to exemplify the aluminum myth, and they indicate that a prospector should not count very heavily upon the value of material discovered by him which, when analyzed, proves to contain things which may have considerable value as elements. Analyses are interesting and informative things, but, when secured, they and samples of the material analyzed should be submitted to a state or federal bureau of mines station or a geological survey before expending any money on the development of the deposit unless, of course, there is absolutely no doubt but that certain elements present can be extracted profitably.

Remarks on Complete Analyses

Many times a year people submit samples to the Arizona Bureau of Mines with requests for complete analyses, and it is probable that other similar organizations have the same experience. It is always necessary to reply that complete analyses are very expensive (one might easily cost $100 or more if the material submitted is at all complex chemically) and are never made except for scientific purposes.

All that a prospector or miner really desires to know is whether something discovered by him contains an element which is present in such a combination and in sufficient quantities so that it can be extracted profitably. A "gray copper" concentrate, for instance, may run 200 ounces to the ton in silver and contain less than one per cent of each of the elements, nickel, cobalt, and lead, but this last fact is of no interest since the presence or absence of the three elements last mentioned has no effect upon the value of the ore. It is important to know the copper and silver content since they are the only elements that can be extracted profitably. The percentage of antimony and arsenic must also be determined since smelters penalize for them if present in considerable quantities. Iron and zinc, which are almost surely present, and other elements may be ignored since almost never present in sufficient quantities to effect the value.

The question that people who have material which they suspect may have value should ask of experts is, then, not "What does a complete analysis show to be present?" but "Is this stuff probably worth anything? If so, for what should it be analyzed?"

In the great majority of instances a few simple tests make it possible to answer these questions with a considerable degree of certainty. The expert will then be able to list the elements of potential value that are apparently present in some quantity and will also list the detrimental elements, if there be any, whose presence or absence should be ascertained. An analysis for all of these elements can then be obtained at a fraction of the cost of a complete analysis.

Of course, the presence or absence of gold and silver can be ascertained only by making an assay, unless recognizable gold or silver minerals are present or gold shows when the material is crushed and panned, but the expert can and will state whether the material is of such a nature that it will possibly prove worth while to assay it. He will not test for platinum and some other very rare elements, however, for he knows that the chance that they are present is almost inconceivably small.

Most states have some department where free tests such as have been mentioned are made and prospectors should utilize the opportunity thus offered to them and not spend time and money developing a deposit of worthless material. They should, however, not expect any state department to spend a week or two making a complete analysis without remuneration and they should not waste money for such analyses.

CHAPTER X

THE DIVINING ROD MYTH[13]

An Ancient and Widely Held Idea

For hundreds, yes, thousands, of years, men have been delving underground for hidden wealth, and probably for the same length of time they have been hoping that some one would invent an applicance that could be relied upon to guide them in their search for minerals and oil. No one knows who first used the forked stick or some other form of divining rod for this purpose, but its lineage is a very ancient, if not an honorable, one. Instruments of this kind have been utilized by thousands of men in probably every country on the globe. Strange as it may seem they are still being used, and a considerable number of the users are men of much better than average intelligence.

Nature of Instruments Used

Divining rods take scores of forms, varying from the simple forked stick or "witch wand," so frequently used in efforts to locate water, to complicated implements elaborately inlaid or filled with various materials. Electric batteries, coils, magnets, etc. sometimes form part of the apparatus which may depart noticeably from the primitive witch-wand form. One type that has been used in Arizona, the oil fields of California, and elsewhere consists of a rubber bulb suspended from a string or wire and filled with mercury. The users of some divining rods claim that their instruments will not work equally well above all kinds of minerals, but must be made receptive or amenable to the influence of whatever metal or substance is being sought by affixing in some way capsules or boxes of mysterious construction and composition. When, for instance, the petroleum capsule is used, the rod is supposed to be attracted only by petroleum; with the gold capsule, it dips only toward gold, etc.

In most instances such devices as are now being considered are held in a certain peculiar way with both hands, and a part of the appliance, so shaped as to make a good pointer, is believed to be strongly drawn toward the material sought. It is, furthermore, sometimes thought to be possible to determine how far such material lies below the end of the instrument. The pointer

---

[13] Part of a paper presented at a Southwestern District, American Association of Engineers convention, held in Las Cruces, New Mexico, April 12, 1922.

on a device manipulated by a person who claims the ability to make such measurements does not turn down toward the object sought and remain at rest, but slightly swings or oscillates up and down before becoming quiescent. The thing sought is then supposed to lie a number of feet below the pointer equal to the number of times this pointer has oscillated. For a Frenchman, of course, the rod obligingly oscillates in terms of meters.

Possibility that Divining Rods May Have Been Used Successfully

Have any previously unknown deposits of mineral or oil been found by manipulators of divining rods? Perhaps. So far as the writer can recollect, he has never heard of a remunerative discovery that was made in this way, but he is perfectly willing to admit that some may have been so made. If an arrow were shot at random into the air in a mineralized district, it might strike the earth above an ore deposit. Similarly the goddess of chance might conceivably smile upon the manipulator of a divining rod and lead him to an ore deposit.

Perhaps it would appear that an idea so hoary with age and so widely accepted must have had its unreliability repeatedly demonstrated, but a little reflection will show that such a conclusion is not necessarily correct. That it means hard luck to break a mirror, to pass under a ladder, or to be one of thirteen persons at a meal are superstitions that have come down to us from remotest antiquity, but no really intelligent, sane person now believes in them, even though current practice tends to keep them alive. If one of thirteen people at a meal dies within a year, the event is noted and widely reported, but no notice is taken of the scores of instances when no fatal results follow such an event.

All Users Not Imposters

Is it, then, true that manipulators of divining rods are to be classed as frauds; do they intentionally deceive those who employ them? In view of what has already been said it may be a surprise to learn that the writer would answer these questions negatively. Considerable investigation has convinced him and others that those who use these implements are, in the majority of instances, honest men who believe in the effectiveness of their work. They do not consciously influence the action of their instruments and they naturally ascribe the phenomena resulting from their use to the operation of supernatural agencies or natural laws that they do not profess to understand or explain. The writer has talked with too many unquestionably honest men who are evidently sincere in their conviction that some favorite form of divining rod operates effectively in their hands, and he has too frequently watched demonstrations staged by such men to doubt their innocence of any desire to deceive.

The pointer does turn down, or the rubber bulb does swing in a certain direction in the hands of such a man; and these effects are not consciously produced by the operator in all, or perhaps most, instances. Furthermore, it is undoubtedly a fact that men operating divining rods have, in many instances, located metals, minerals, etc., even when wholly ignorant of their positions. provided some one else knew where they were.

Facts Bearing on the Solution of the Problem

The question of why divining rods do sometimes operate in the hands of honest men is certainly a puzzling one, but its solution does not seem to be impossible. Before attacking the problem several facts should be known, namely:

1. Excepting for dipping needles used in locating iron ores and other types of geophysical apparatus discussed later, which are scientific instruments that embody no mysterious features. no form of divining rod will operate unless held in human hands.

2. Nearly all known forms of such implements can be forced to operate by conscious muscular effort even when known to be far from any substance supposed to attract them, and their action under such conditions may be accompanied by all the phenomena that follow their use when seeking the unknown position of such substances. Among such manifestations are the twisting of green bark away from the wood beneath, the bobbing of the pointer to indicate the depth, etc.

3. If a person believes that a certain type of divining rod is the only one that can be satisfactorily used, no other variety will operate effectively for him. If he thinks that a rod will not be attracted by gold unless fitted with a certain attachment, it will not dip toward gold for him without this modification, but, in the hands of another person who does not believe in the use of such attachments, it may seem to work equally well with or without the thing that the first man thinks makes it sensitive to gold.

Careful investigation indicates that all types of divining rods to which the above statements do not apply are probably operated by people who realize their uselessness and are not averse to deceiving others.

Theories Explaining the Action of Divining Rods

It must be evident that there can be only four possible explanations of the operation of divining rods, namely:
  1. Conscious muscular effort exerted by the holder.
  2. Supernatural agencies.
  3. The operation of unknown and mysterious natural forces.
  4. The unconscious or involuntary muscular efforts of the manipulator.

1. The first suggested solution of the mystery needs no discussion. While it doubtless accounts for the actions of the instruments used by people who seek to prey upon the gullible, it does not explain the phenomena observed when the devices are used by honest men.

2. If divining rods are operated by supernatural agencies, we must assume that an American or English spirit directs the instrument of an American when he computes the depth of an ore body in feet by noting the number of times the pointer oscillates, and that a spirit familiar with the metric system is the directing force behind the Frenchman's rod when he measures depths in meters. We must further suppose that an absolutely intangible, immaterial being is able to pull hard enough on the pointer to twist the bark off of green wood. Possibly, however, the spirit merely causes the receptive human agent's subsconscious mind to direct the muscles to flex in such a way as to make the divining rod operate. If a spirit can at will impress an idea upon a human mind in this way, we must admit, first, that spirits seem to take a rather strange interest in buried substances of more or less intrinsic value to men; second, they show a commendable interest in helping men find them; third, they fail so frequently that their omniscience is open to grave doubt; and, fourth, they show an odd willingness to conform in their activities to the peculiar and varied beliefs and notions of the men whom they try to help.

It is believed that all intelligent people will find it impossible to harmonize their ideas of the existence of a soul after death with such absurd activities on its part, and intelligent people no longer believe in demons.

3. Unknown natural forces of some kind, usually supposed to be electrical or magnetic, do not constitute an adequate explanation of the action of divining rods in the minds of scientists—men who have devoted their lives to the study of natural forces of all kinds. The diversity in the form and construction of the many different divining rods used and their contradictory behavior in the hands of many individuals seem utterly incompatible with the theory that they are operated through the action of such natural forces.

4. Many painstaking investigators and observers now feel perfectly sure that the last suggested explanation is the proper one, i.e., divining rods are operated by the involuntary flexing of the muscles of the manipulator. It is necessary, however, to go into the matter a little more fully in order to explain how material that has been hidden or is known to exist at a certain place by someone can be located by another who has no knowledge of its presence. That some manipulators of divining rods can do feats of this kind is generally admitted.

The writer hesitates to mention the only explanation that occurs to him because it lies in the realm of a comparatively new and still imperfectly understood field of investigation. He suggests, however, that thought transference may cause the subconscious mind to flex the muscles involuntarily in such a way as to make the pointer dip and even to oscillate in the hands of some people.

It appears to be certain that the operation of divining rods constitutes a very promising and attractive field of research for psychologists, and it is to be hoped that someone with the requisite training and interest will soon investigate the whole subject.

The suggestion that thought transference plays an active part in the operation of these implements appears to be a likely explanation of the phenomenon when it is remembered that they seem to work best when someone in the party knows the position of the thing sought, a condition that frequently exists when the manipulator of a divining rod is called upon to demonstrate his powers in a well developed mineralized area with which at least one man in the party is thoroughly familiar, but concerning which the manipulator knows absolutely nothing. The rather astonishing successes that sometimes follow such efforts lead naturally to the conclusion that the apparatus used will be equally effective when employed in virgin territory. The manipulator is himself deceived, and his failure to give useful information when seeking to locate undiscovered deposits is as much of a surprise and disappointment to him as to those who employ him.

In response to the question of why divining rods operate at all in places where none of the substance sought exists and where no mental suggestions could be received from anyone else, it can only be suggested that the subconscious mind of the operator plays tricks on him. Unconsciously he observes certain surface features that his experience has led him to associate with deposits of the substance sought, and the action of his instrument is an unconscious response to such suggestions.

The writer is fortunate enough to know one man who was once a thorough believer in divining rods. He even convinced a minister of the gospel that a certain type of instrument would always operate in the hands of either of them when in the vicinity of a producing oil well. The minister engaged in the business of locating unknown deposits of petroleum by the use of the device, but my friend's later experience convinced him absolutely that divining rods could never be depended upon to locate previously undiscovered deposits. He diligently tried to convince the reverend gentleman mentioned that he was unconsciously victimizing people who employed him, but was not successful.

Some people believe that it is not even necessary that a person who knows the situation of a deposit be in the company of a man operating a divining rod; they claim that, if anyone, anywhere, has information concerning the deposit, it is not necessary that this person shall be with the divining rod manipulator in order that his instrument shall indicate the position of the deposit. Statements even more difficult to believe than this one are sometimes made, but the writer prefers to withhold judgment and comment until scientific experiments (which should not be difficult to make) have demonstrated their truth or falsity.

No matter why divining rods operate in the hands of honest men, there is no doubt of their utter unreliability as locators of undiscovered substances. If they were worth anything, practically all of the deposits of mineral and oil, at least in the more populous countries, would long ago have been located. In prospecting as in most other lines of endeavor, there are few short cuts to success. When prospecting in virgin territory little reliance can be placed on anything but the eye and brain, and most outcrops of mineral deposits will be found in the future, as in the past, by vigorous walkers and climbers who seek and find "float," and trace it to its source, or who carefully examine all existing rock exposures in a given area.

Divining Rods and Geophysical Instruments Differentiated

Within the last fifteen years a new science, geophysics, has been developed. As the word is now generally used by mining men it relates to the location of ore bodies, petroleum deposits, etc., beneath the earth's surface by means of scientific instruments in no way related to divining rods since the forces acting upon the geophysical instruments are perfectly understood and the results secured are in no way mysterious to anyone who has studied and mastered the operation of these instruments.

Classification of Geophysical Methods

Four different geophysical methods have been developed which may be termed (1) the gravitational, (2) the seismic, (3) the magnetic, and (4) the electrical.

The Gravitational Method.—This method is based on the fact that the force of gravity at any point on the earth's surface, which is measured by an instrument called a torsion balance, is affected by the density of the earth materials underlying that point. Large deposits of relatively light material, like the rock salt in salt domes, can be easily located with a torsion balance.

The Seismic Method.—It has long been known that earthquake waves travel with different velocities through the varied substances of which the earth's crust is composed. When the seismic method is used an artificial earthquake is created by firing a heavy charge of some explosive, and the velocity with which the earth waves travel through surrounding material is determined. Many useful facts concerning the nature of the earth materials traversed by the waves may thus be obtained.

The Magnetic Method.—Variations in the magnetic attraction of the earth are measured when this method is used. The instrument employed is called a magnetometer and a magnetic dipping needle is a very simple form of such an instrument.

The Electrical Method.—Three different electrical methods, namely, the inductive, the self-potential, and the induced potential, are in general use. It would serve no good purpose to attempt to distinguish between them or to describe them in detail in a book like this one, but it may be said that, in all of them, delicate instruments are used either to detect the presence and direction of flow of electrical currents that are set up when an ore body oxidizes, as in the self-potential method, or to ascertain the effect of an ore body upon the direction of flow of artificially produced electrical currents which are caused to pass through the earth, as is done in the other two methods mentioned.

Practical Value of Geophysics

The first three methods have been used very successfully in the location of deposits of petroleum and natural gas. According to J. Brian Eby[14] more than sixty salt domes were located by the use of these methods on the Gulf coast of Texas and in Louisiana between 1923 and 1931, and there were only forty-six such fields known in the coastal region prior to 1923 when geophysical methods were first used in locating petroleum in the area mentioned.

While the gravitational and seismic methods have been used very successfully by people who are searching for petroleum, they have not been useful in the location of ore deposits. Geophysical work on ore deposits is carried on principally by the use of one or more of the electrical methods, although a magnetometer is found useful in some instances.

A well trained, experienced man can, by the use of electric geophysical methods, often locate deposits of sulphide ore beneath barren or oxidized outcrops or undisturbed soil and other overburden, and he can state approximately how deep the top of the ore body is below the surface. If the deposit is in a vein he can obtain some information concerning its strike and dip. Except that it is a metallic ore he cannot state its nature or grade, but he can often save people who are developing properties a great deal of money, since he can indicate the areas that are barren or potentially ore bearing.

Although small, portable electrical instruments have been devised for the use of prospectors, they are not powerful enough to be affected by anything but very rich pockets or large sulphide ore bodies relatively close to the surface, and their usefulness is very limited.

Long and thorough study of physics and mathematics and considerable experience are required to enable a person to build and use geophysical instruments satisfactorily, so most miners must employ others to do such work for them. The work of the geophysicist is a highly specialized calling and there are now relatively few men who are qualified to follow it satisfactorily. The qualifications of people offering to make geophysical surveys should, then, be very carefully scrutinized before they are employed.

---

[14] "Relation of Geology to Geophysics on the Gulf Coast," Bul. Geol. Soc. of Amer., March, 1932, p. 249.

APPENDIX A

BRIEF DESCRIPTIONS OF MINERALS MENTIONED IN THIS BOOK

Adularia

Silicate of aluminum and potassium. A variety of orthoclase feldspar that is usually white or colorless and has a glassy luster; too hard to be scratched with a knife; breaks smoothly in two directions at right angles to each other; sometimes occurs as gangue mineral, but is valueless.

Aegirite

Silicate of sodium and iron. A mineral that is not easily identified without microscopic work, but it occurs in hard, dark colored, glassy, needle-like crystals in igneous rocks that are usually light in color and have formed on or relatively close to the surface of the earth, such as rhyolites, trachytes, and phonolites.

Alabandite

Sulphide of manganese. Easily scratched with a knife; dull metallic luster; black, but may have a brown tarnish; dark green powder; breaks in three directions at right angles to each other; an ore of manganese that contains 63.1 per cent of that metal when pure, and is found below the oxidized zone.

Alunite

Hydrous sulphate of aluminum and potassium. Easily scratched with a knife; usually in dull lustered, white, faintly flesh tinted, or grayish, finely granular and sometimes porous masses, but may be found in glassy white or colorless crystals; not easily identifiable without chemical or blowpipe tests; sometimes occurs as a gangue mineral and is a commercial source of potash.

Amethyst

A purple variety of glassy quartz (which see); a semiprecious gem, but an American amethyst has little value since it fades quickly.

Anglesite

Sulphate of lead. Easily scratched with a knife; crystals and crystalline masses have an adamantine luster or, in other words, reflect light so perfectly that surfaces appear to be covered with thin, transparent films of some silvery metal; white or colorless when in crystals and, usually, white, gray, brownish, or yellowish, and dull or resinous lustered when massive; surfaces of the massive material often show dark rectangular markings; very heavy; a valuable ore of lead that contains 73.6 per cent of that metal when pure and is confined to the oxidized zone; it often carries silver.

79

Ankerite

Carbonate of calcium, mangesium, and iron, with usually a little manganese. Easily scratched with a knife; stony, glassy, or pearly luster; white, gray, brownish. or reddish; breaks smoothly in three directions at broad or acute angles; dissolves quickly in muriatic acid with the emission of numerous bubbles; occurs coarsely to finely granular, compact, and in crystals which often have warped faces; a valueless gangue mineral.

Apatite

Phosphate of calcium with chlorine and fluorine. Can be scratched with a knife if considerable pressure is applied; typically glassy luster, but some varieties are dull and earthy; green, brown, or yellow when glassy, or white, yellowish, or grayish when dull; breaks with a curving or rough fracture; sometimes occurs in six-sided crystals; the dull, earthy variety cannot be identified without chemical or blowpipe tests; after treatment, it is a valuable fertilizer.

Aquamarine

A light blue variety of beryl (which see). A valuable semiprecious gem, the value increasing with the depth of the color.

Argentite (Silver Glance)

Sulphide of silver. Soft, and, when pure, it may be cut like a piece of hard lead; metallic luster; lead-gray to black; lead-gray powder; rough fracture; a valuable ore of silver and contains 87.1 per cent of that metal when pure.

Arsenopyrite (Arsenical Pyrites or Mispickle)

Sulphide and arsenide of iron; too hard to be scratched with a knife; metallic luster; silver-white to slightly grayish, but may have a yellowish tarnish; dark gray powder; coarsely crystalline material breaks smoothly in two directions at angles of 112 degrees or 68 degrees; occurs in coarsely to finely granular masses, disseminated grains, and crystals; when struck a glancing blow with a hammer, emits sparks and an odor of garlic which will distinguish it from all other minerals excepting cobaltite and smaltite which are much rarer; the most important source of arsenic and may contain gold.

Asbestos

This name is applied to the fibrous variety of several different minerals of which the two commonest and most important are chrysotile, a variety of serpentine (a hydrous silicate of mangesium and iron), and amphibole asbestos (anhydrous silicate of numerous bases); chrysotile asbestos is often very flexible and tough, while amphibole asbestos is usually more brittle.

Azurite

Hydrous carbonate of copper. Soft enough to be easily scratched with a knife; glassy to dull luster; dark blue when in crystals and light blue when earthy; blue powder; dissolves quickly in muriatic acid with the emission of numerous bubbles and the greenish solution thus formed becomes dark blue when ammonia is added; a common ore of copper that contains 55.4 per cent of that metal when pure and is found only in the oxidized zone.

Barite (Barytes or Heavy Spar)

Sulphate of barium. Soft enough to be easily scratched with a knife; glassy to stony luster; usually white or light tints of yellow, brown, gray, or red; white powder; breaks easily and smoothly in one direction and less easily and smoothly in two other directions which make an angle of about 101 degrees with each other and are both perpendicular to the perfect cleavage first mentioned; transparent to translucent; about 70 per cent heavier than quartz and its weight is its most characteristic property; occurs in crystals and in platy and granular masses; has several industrial uses and there is considerable demand for pure material; a common gangue of lead and silver ore.

Bauxite

Hydrous oxide of aluminum. Soft enough to be easily scratched with a knife, and sometimes, with a finger nail; dull luster; white, yellow, brown, or red; color of powder is same as color of mineral; has odor of clay when moistened; broken surfaces often show dark round spots or rings, a fraction of an inch across, in a lighter, clay-like groundmass; practically the only ore of aluminum and contains 39 per cent of that metal when pure, but can rarely be recognized with certainty without chemical tests.

Beryl

Silicate of beryllium and aluminum. Resembles quartz (which see) so closely that the two minerals often cannot be distinguished at sight; is commonly bluish or greenish in color which tints quartz rarely shows; often found in six-sided crystals in pegmatite dikes, and these crystals differ from quartz crystals (also six-sided) in that they are usually terminated at the end by a single face while quartz crystals are terminated by either three or six faces that meet at a point, and, also, the six prismatic faces on quartz crystals show fine lines or markings which run horizontally around a crystal when it is held upright, while beryl crystals when held in the same position may show vertical, but not horizontal, lines; may be further distinguished from quartz by the fact that quartz may be scratched with a piece of beryl while beryl cannot be scratched with quartz; beryl is the only source of the valuable, light metal beryllium, and contains 5.1 per cent of that metal when pure; emerald (green), aquamarine (light blue), and morganite (pink) are valuable gem varieties.

Bismuth

Native bismuth which usually contains traces of arsenic, sulphur, and tellurium. Easily scratched with a knife, and sometimes with the finger nail; metallic luster; silver-white with a slightly reddish hue, but is subject to tarnish; powder same color as untarnished mineral; brittle, but may be flattened somewhat without breaking when heated; breaks easily and smoothly in one direction and less so in one or two other directions; usually occurs in tree-like groupings of indistinct crystals and in platy or granular masses; a very rare mineral.

Bismuthinite

Sulphide of bismuth. Soft enough to be easily scratched with the finger nail; metallic luster; color of both mineral and powder is silver-white or slightly darker, but the mineral commonly shows a yellowish or iridescent tarnish; commonly associated with chalcopyrite and pyrite or bismutosphaerite, a yellow or brown earthy carbonate of bismuth; an ore of bismuth that contains 40.6 per cent of that element when pure, but it is a relatively rare mineral.

Bornite (Peacock Ore)

Sulphide of copper and iron. Soft enough to be scratched with a knife; metallic luster; reddish brown, but quickly tarnishes iridescent; dark gray powder; very brittle; lighter than iron; usually associated with chalcopyrite; a valuable ore of copper and may contain as much as 55 per cent of that metal.

Calamine

Silicate of zinc. May be scratched with a knife; glassy luster; usually colorless or white, but may be tinted light shades of yellow or brown or, more rarely, green; white powder; occurs usually in granular masses or as groups of numerous, flat crystals that are united into coxcomb-like groups and show a coarsely radiating structure when broken; a relatively uncommon oxidized ore of zinc that contains 54.1 per cent of that metal when pure.

Calcite

Carbonate of calcium or lime. Very easily scratched with a knife; glassy to dull luster; colorless or white when pure and pale to deep shades of many colors, even black, when impure; white powder; breaks very easily and smoothly in three directions which make angles of about 105 degrees or 75 degrees with each other; dissolves quickly in either concentrated or dilute, hot or cold muriatic acid with the emission of numerous bubbles; occurs in six-sided crystals, granular masses (marble), compact, dull lustered masses (limestone), porous and earthy masses (chalk), etc.; has many industrial uses as, for instance, the manufacture of cement and quicklime.

Cassiterite

Oxide of tin. Too hard to be scratched with a knife; glassy to dull luster; usually brown, yellow, reddish, or black; very light brown powder; nearly as heavy as iron; occurs in four- or eight-sided crystals, granular to earthy masses, disseminated grains, layers with surfaces that resemble a bunch of grapes or kidney, and water-worn pebbles made up of layers that may be composed of radiating fibers; the principal ore of tin and contains 78.6 per cent of that metal when pure.

Celestite

Sulphate of strontium. Much like barite (which see), but is slightly lighter in weight, is apt to occur in marble-like, granular masses with a light bluish color, and is not gangue mineral; a source of the strontium nitrate used in red fireworks and signals.

Cerussite

Carbonate of lead. Closely resembles anglesite (which see) in every way except that it will dissolve with the emission of numerous bubbles in hot concentrated or boiling dilute muriatic acid; contains 77.7 per cent of lead when pure and is an oxidized ore of that metal.

Chalcedony

Oxide of silicon. A variety of quartz (which see) that is composed of such excessively minute particles that they may be seen only with a high-powered microscope; waxy to dull luster; breaks with a very smooth, curving fracture; usually pale blue, gray, or drab in color; valueless although some varieties are cut and erroneously called moonstone.

Chalcocite (Copper Glance)

Sulphide of copper. When pure may be easily cut with a knife, leaving a polished surface; metallic luster or dull when tarnished or in the form of powder; dark lead-gray or black when tarnished or in the form of powder; lead-gray powder unless the mineral itself occurs as a black powder; smooth, curving fracture; usually associated with other ores of copper of which it is one of the most valuable since it contains 79.8 per cent of that metal when pure and often carries silver; found usually in the deeper parts of copper deposits that have not been affected by atmospheric gases carried downward in rain water.

Chalcopyrite (Fools' Gold)

Sulphide of copper and iron. Soft enough to be scratched with a knife; metallic luster; bright brass yellow, but often has an iridescent tarnish; greenish black powder; uneven fracture; very brittle; when pure contains 34.5 per cent of copper and is one of the most important ores of that metal; may carry some gold and silver; found in the deeper parts of copper deposits that have not been affected by atmospheric gases carried downward in rain water.

Chromite

Oxide of chromium and iron. Too hard to be scratched with a knife; luster varies from that of black glass to metallic or dull; black; dark brown, sometimes reddish powder; when it looks like black glass, each grain has a very smooth, curving fracture, but the fracture is usually uneven; often associated with serpentine (which see); may be coated with bright green zaratite, a hydrated basic carbonate of nickel; principal source of chromium and contains 46.2 per cent of that metal when pure.

Chrysocolla

Hydrous silicate of copper. Soft enough to be scratched with a knife and may sometimes be scratched with a finger nail; somewhat glassy to dull luster; blue to greenish blue and green, and is sometimes spotted with brown; white powder; smooth, curving fracture; most specimens adhere strongly to a nearly dry tongue; very light in weight; often associated with other copper minerals in deposits relatively close to the surface; an unimportant ore of copper although it contains 45.2 per cent of that metal when pure; siliceous chrysocolla is used as a semiprecious gem; a black variety of this mineral is called copper pitch ore.

Cinnabar (Native Vermilion)

Sulphide of mercury (quicksilver). Very easily scratched with a knife and, often, with a finger nail; luster varies from that described as shown by crystals of anglesite (which see) to dull; bright red in crystals, dark red to reddish black when in compact masses, and bright red when earthy; scarlet to brownish red powder; somewhat heavier than iron; rarely associated with any other than gangue minerals or pyrite; if the powder is moistened with muriatic acid and rubbed on a bright copper coin, the coin will be covered with mercury; the only important ore of mercury and contains 86.2 per cent of that element when pure.

Cobaltite

Sulphide and arsenide of cobalt. Too hard to be scratched with a knife; metallic luster; tin-white to steel-gray, but often has a slight pink or violet tint; black powder; often associated with erythrite, an earthy, pink arsenate of cobalt; yields sparks and the odor test described under arsenopyrite (which see), but may usually be distinguished therefrom by the color; a valuable ore of cobalt which contains 35.5 per cent of that metal when pure and may carry silver.

Columbite (Niobite)

Columbate of iron and manganese. Too hard to be scratched with a knife; nearly metallic to somewhat resinous luster; black to brown; often has an iridescent tarnish; dark reddish brown powder; breaks in two directions at right angles to each other; nearly as heavy as iron and heavier than magnetic iron oxide; a very rare mineral that is usually found in pegmatites.

Corundum

Oxide of aluminum. Easily scratches any other mineral except diamond; glassy to stony luster; all colors, but usually gray when massive; white powder; often breaks or parts smoothly in three directions at nearly right angles to each other, and parting faces are finely cross-hatched with lines at nearly right angles; about 50 per cent heavier than quartz; often occurs in six-sided crystals; when mixed with magnetite and other impurities, it is called emery; transparent, flawless material constitutes the very valuable gems ruby (red), sapphire (blue), etc.

Copper

Native copper, usually with some silver and gold. Easily cut with a knife; metallic luster; copper-red, but easily tarnishes green or black; may be pounded into thin sheets without breaking; heavier than iron; occurs in grains, sheets, irregular masses, and crystals that are usually distorted.

Cryolite

Fluoride of sodium and aluminum. Very easily scratched with a knife; somewhat greasy luster; usually snow-white or colorless, but sometimes reddish or brownish; white powder; breaks roughly in three directions at nearly right angles; often associated with small amounts of siderite and galena; the mineral first used as a source of aluminum (of which it contains 12.8 per cent when pure) and is still used as a flux in the extraction of that metal from bauxite.

Cuprite (Ruby Copper)

Oxide of copper. Easily scratched with a knife; the luster of crystals is like that of crystallized anglesite (which see), but noncrystallized material may be glassy to somewhat metallic or dull; various shades of red and sometimes nearly black; red powder; over twice as heavy as quartz and heavier than magnetic iron oxide; occurs massive, earthy, and as octahedral crystals which are sometimes elongated until as slender as hairs (chalcotrichite); found only in the oxidized zone; contains 88.8 per cent of copper when pure.

Diamond

Native carbon. The hardest mineral—can be scratched only with diamond; luster like that described for crystals of anglesite (which see); white, brownish, yellowish, greenish, and pale tints of any color; splits or cleaves easily and smoothly parallel to the faces of an octahedron—in four directions at angles of about 72 degrees and 108 degrees; a third heavier than quartz; usually occurs in rounded octahedrons, and, much less commonly, cubes; also occurs as bort or carbonado with a confused crystalline texture, no cleavage, and sometimes decided porosity; found in very basic igneous rock (peridotite) and placers.

Emerald

A green variety of beryl (which see), free from any yellowish or bluish tint.

Enargite

Sulpharsenate of copper. Easily scratched with a knife; bright metallic luster; black; black powder; splits easily and smoothly in two directions at angle of about 98 degrees; occurs in columnar masses and in lustrous crystals; an ore of copper that contains 48.3 per cent of that metal when pure.

Epidote

Silicate of calcium, aluminum, and iron. Too hard to be scratched with a knife; glassy to stony luster; dark green to greenish black when in crystals and light yellowish green when finely granular; white powder; crystalline material splits easily and smoothly in one direction; occurs in long, brilliant, deeply grooved crystals, cleavable masses, and finely granular; a common, valueless mineral which is often erroneously supposed to be a copper ore.

Fluorite

Fluoride of calcium. Soft enough to be scratched with a knife; usually glassy; green, violet, pink, purple, yellow, brown, or white; white powder; splits or cleaves exactly like diamond (which see); occurs in cubical crystals, cleavable masses, and coarsely to finely granular; a common gangue mineral that occurs with lead and silver ores and it has several uses which give it some value.

Galenite (Galena)

Sulphide of lead. Easily scratched with a knife; bright metallic luster; lead-gray, sometimes tarnished black; lead-gray powder; splits very easily and smoothly parallel to the faces of a cube—in three directions at right angles to each other; very brittle; heavier than iron; occurs in cubo-octahedral crystals, cleavable masses, and coarsely to finely granular; the principal ore of lead, containing 86.6 per cent of that metal when pure, and often carrying considerable silver, particularly when finely granular.

Garnet

A silicate that may contain many different bases, depending upon the variety; calcium, aluminum, and iron are common ingredients. About as hard as quartz; glassy to resinous luster; usually some shade of red, brown, or yellow; white powder; fracture is usually uneven; sometimes noticeably heavy; often found in spheroidal crystals and in large masses formed by the action of heat and mineralizing solutions on impure limestone; transparent, attractively colored material constitutes a semi-precious gem and certain of the harder varieties that break with sharp edges form a valuable abrasive.

Gold

Native gold with some silver and other impurities. Easily cut with a knife without crushing or yielding powder; golden yellow to nearly silver-white and does not tarnish; may be pounded into thin sheets without breaking; about twice as heavy as iron; occurs in grains, nuggets, wires, sheets, distorted crystals, etc.

Graphite (Plumbago or Black Lead)

Native carbon. Easily scratched with the finger nail; metallic to dull luster; dark gray to black; gray powder and makes gray mark on glazed porcelain or paper; metallic lustered material easily splits into thin, flexible flakes; feels greasy and soils fingers; occurs in scales, platy to compact masses, and, rarely, as six-sided, flat crystals; used in the manufacture of lead pencils, lubricants, crucibles, etc.

Gray Copper

A term applied to both tetrahedrite and tennantite (which see).

Halite (Rock Salt)

Chloride of sodium. Very easily scratched with a knife; glassy luster; colorless or white when pure, but may be tinted yellow, brown, red, etc. by impurities; white powder; splits very easily and smoothly parallel to the faces of the cube—in three directions, at right angles to each other; is common table salt and has the distinctive taste of that substance; occurs in cleavable granular or compact masses and as crystals.

Hematite

Oxide of iron. Too hard to be scratched with a knife except when earthy or finely micaceous; metallic to earthy luster; dark steel gray, reddish gray, or red; red to dark red powder; often has a platy texture; occurs in micaceous masses, or thin flakes (specular iron), or in crystals, all with a metallic luster, also in compact, kidney-shaped, and oolitic (composed of tiny spheroidal grains) masses with a dull luster, and in hard (ruddle or red ocher) to soft (red ocher) earthy masses; the principal ore of iron, and is used for other purposes, such as cheap paint; if the finely micaceous material (which is commonly confused with some valuable sulphide) is hit vigorously against the palm of one's hand, the skin will be covered with tiny, glittering scales, and no other mineral yields this test.

Horn Silver

A term applied to all compounds of chlorine, bromine, and iodine with silver. The commonest such compound is cerargyrite (chloride of silver) and most of the other horn silver minerals resemble it so closely in appearance and value that it is not necessary to distinguish between them. Cerargyrite has the following characteristics: cuts like wax and it easily scratched with a finger nail; waxy to resinous luster; light gray, greenish, or yellowish when freshly mined, but may become brown on exposure to the light; white powder; if rubbed on moistened zinc or iron, it swells, turns black, and is reduced to spongy metallic silver; occurs usually as crusts or veinlets, but sometimes as masses or crystals; a very valuable ore of silver, which is found only in the oxidized zone and contains 75.3 per cent of that metal when pure.

Hydrozincite (Zinc Bloom)

Hydrous carbonate of zinc. Often soft enough to be scratched with a finger nail; dull luster; white, grayish, or yellowish; white powder; often noticeably heavy; dissolves quickly in muriatic acid with the emission of numerous bubbles; usually occurs as chalklike masses which are often associated with smithsonite or other zinc ores as an alteration product in the oxidized zone.

Ilmenite

Oxide of iron and titanium. Usually too hard to be scratched with a knife although some varieties may be thus scratched when great pressure is applied; dull metallic luster; iron-black; brownish black to reddish brown powder; small surfaces may show a smoothly curved but rather flat fracture; feebly attracts a magnetic needle; occurs in compact, granular masses, thin plates, imbedded grains, sand, and, rarely, crystals; in some demand as a source of titanium.

Kaolin (Kaolinite)

Kaolinite is a mineral of definite properties and composition (hydrous silicate of aluminum), while kaolin is a term used for all soft, earthy-lustered, white or light tinted, commonly plastic material that feels smooth and has a strong, earthy odor when moistened. It is the principal constituent of clays.

Leucite

Silicate of potassium and aluminum. Too hard to be scratched with a knife; the crystals are dull on the surface and have a greasy luster in the interior; white to light gray; white powder; occurs as grains or spheroidal crystals or groups of crystals imbedded in volcanic rocks that have long been used as mill stones; crystals often have octagonal cross sections; a rather unimportant source of potash.

Limonite

Hydrous oxide of iron. Some varieties are too hard to be scratched with a knife, but earthy limonite may appear to be very soft; luster usually dull; yellow and brown to nearly black; yellow to yellowish brown powder; occurs in earthy to compact masses, crusts with radiating texture and sometimes varnish-like surfaces, stalactites, etc.; the commonest ore of iron and is used as a cheap yellow or brown paint (yellow ocher or umber); found only in the oxidized zone.

Magnesite

Carbonate of magnesium. May be scratched with a knife although noticeably harder than calcite (which see) which it may otherwise resemble closely; earthy to glassy luster; white, yellowish, or grayish; white powder; the earthy, massive material has a smoothly rounded fracture while crystalline material splits easily and smoothly in three directions at angles of about 73 degrees and 107 degrees; dissolves and emits numerous bubbles when placed in hot concentrated muriatic acid, but is unaffected by cold acid; occurs in white, earthy masses like hard chalk, in cleavable, marblelike masses, and, rarely, in crystals which are sometimes black in color due to included organic material; an important source of magnesia.

Magnetite

Oxide of iron. Too hard to be scratched with a knife, metallic to rather dull luster; black; black powder; occasionally parts or splits parallel to the faces of an octahedron; strongly attracted by a magnet and is itself sometimes a magnet (lodestone); occurs compact and platy massive, coarsely to finely granular, as sand (it is the principal ingredient of black sand), and in octahedral crystals; an ore of iron.

Malachite

Hydrous carbonate of copper. Easily scratched with a knife; glassy, silky, or dull luster; bright emerald green to dark green; green powder, paler than the color; dissolves easily when placed in muriatic acid, with the emission of numerous bubbles, and the green solution then formed becomes dark blue when ammonia is added to it; occurs in crusts with a radiating texture, fibrous or banded compact masses, granular or earthy, and in slender crystals; a common ore of copper, found only in the oxidized zone, and contains 57.5 per cent of that metal when pure.

Marcasite

Sulphide of iron. Like pyrite (which see), but is lighter in color (pale brass yellow), the crystals are not cubical, and it is formed only at comparatively shallow depths.

Massicot

Oxide of lead. Easily scratched with a finger nail; dull luster; sulphur-yellow or slightly orange tinted; light yellow powder; occurs earthy or somewhat scaly; contains 92.8 per cent lead when pure, but is usually impure; confined to the oxidized zone.

Mica

A general term used to cover a number of micaceous minerals of which muscovite or white mica (which see) is the only species described herein.

Mimetite

Arsenate of lead with chlorine. Cannot be distinguished with certainty from pyromorphite (which see) without using blowpipe or chemical tests, but the color is usually yellow or brown, and more rarely white, instead of greenish; an unimportant ore of lead, confined to the oxidized zone, which contains 69.7 per cent of that metal when pure.

Minium

Oxide of lead. Very soft and can often be scratched with a finger nail; dull luster; bright red color although often mixed with some yellowish material; orange-yellow powder; occurs as an earthy or scaly powder; contains 90.6 per cent lead when pure, and is confined to the oxidized zone.

Molybdenite

Sulphide of molybdenum. Its properties are almost identical with those of graphite (which see) excepting that it leaves a light yellowish green mark when rubbed on glazed porcelain or paper; the principal source of molybdenum, used in steel, dyes, etc., and contains 60' per cent of that metal when pure.

Muscovite

Hydrous silicate of potassium and aluminum. Soft enough to be scratched with a finger nail; glassy to pearly luster; colorless, white, or light tints of green, brown, or yellow; splits very easily and smoothly into extremely thin sheets that are both flexible and elastic; very tough; occurs in plates, scales, micaceous masses, and six-sided crystals; used in making roofing, spangling wall paper, in the electrical industry, etc., and large sheets of flawless, colorless material are valuable.

Opal

Hydrous oxide of silicon. May usually be scratched with a knife if extreme pressure is applied; greasy to glassy luster; all colors, but commonly white; breaks with a very smooth, polished, curving fracture; occurs in veinlets and filling cavities relatively near the surface; if it shows internal reflections of various colors it is precious opal, a gem-stone.

Orthoclase

Silicate of potassium and aluminum. Too hard to be scratched with a knife, but can be scratched with quartz; stony to glassy luster; usually flesh-colored, white, or yellow, but may be gray, brown, red, or, rarely, green; white powder; splits easily and smoothly in two directions at right angles to each other; usually associated with quartz and mica; occurs in cleavable masses, grains, and crystals; used in the manufacture of glass and porcelain.

Peridot (Olivine or Chrysolite)

This term is applied to yellowish green, transparent chrysolite that is a semiprecious gem. The characteristics of chrysolite or olivine are as follows: silicate of magnesium and iron; too hard to be scratched with a knife; glassy luster; dark bottle green to golden yellow or brown; white powder; smoothly curving fracture; occurs in granular masses, grains, or crystals imbedded in basic igneous rocks, such as basalt.

Plumbojarosite

Sulphate of lead and iron. Easily scratched with a knife; somewhat glassy to dull luster; dark brown; light brown to yellow powder; us-ually occurs in minute, tabular crystals; a difficult mineral to recognize without chemical tests.

Polybasite

Sulphantimonite of silver and copper. Very easily scratched with a knife; metallic luster; iron black, but is sometimes cherry-red in thin splinters; black powder; very brittle; uneven fracture; about as heavy as iron; usually occurs in six-sided crystals with beveled edges; a rare but valuable ore of silver that contains 75.6 per cent of that element when pure.

Proustite (Light Ruby Silver)

Sulpharsenite of silver. Very easily scratched with a knife; luster like that of crystals of anglesite (which see) to almost metallic; ruby red to almost black and darkens on exposure to the light; scarlet powder; uneven to smoothly curving fracture; occurs as disseminated grains, crusts, veinlets, compact masses, and six-sided crystals; usually associated with other silver minerals; it is a valuable ore of silver and contains 65.4 per cent of that metal when pure.

Psilomelane

Impure, hydrous oxide of manganese. Too hard to be scratched with a knife except when in the form of wad or bog manganese which is earthy, has a brown powder, and is soft enough to soil the fingers; semimetallic to dull luster; iron-black; dark gray powder; very smooth, curving fracture; occurs in crusts that have spheroidal projections all over the surface, flint-like masses, or in layers with pyrolusite (which see); the commonest ore of manganese, but is not so high-grade or pure as pyrolusite.

Pyrargyrite (Dark Ruby Silver)

Sulphantimonite of silver. Like proustite (which see), but is usually black in color although thin splinters may be ruby red by transmitted light; it has a dark purplish red powder, and it contains only 59.9 per cent of silver when pure.

Pyrite (Pyrites or Iron Pyrites)

Sulphide of iron. Too hard to be scratched with a knife; bright metallic powder; pale to full brass yellow and sometimes almost silver-white; black powder; uneven to smoothly curving fracture; occurs in compact or granular masses, as disseminated grains, and, frequently, in well-formed cubical, octahedral, or spheroidal crystals; the commonest sulphide mineral and may, rarely, contain enough gold to constitute an ore of that metal; also used in the manufacture of sulphuric acid.

Pyrolusite

Oxide of manganese. Usually soft enough to be scratched with a finger nail and soils the fingers; metallic to dull luster; black to steel-gray; black powder, often sooty; occurs in indistinct crystals, radiating needles, velvety crusts, and granular masses, and is sometimes interlayered with psilomelane; the principal ore of manganese and contains 63.2 per cent of that metal when pure.

Pyrrhotite

Sulphide of iron. Easily scratched with a knife; metallic luster; light brown; dark gray powder; uneven fracture; weakly to strongly magnetic; usually occurs in granular masses or disseminated grains; may contain enough nickel to constitute an ore of that metal.

Quartz (Silica)

Oxide of silicon. Too hard to be scratched with a knife; if finely granular like chalcedony (which see), agate, etc., may have a dull luster; but most other varieties, of which there are many, mostly based on color, have a glassy luster; most commonly white or colorless, but sometimes yellow (rusty quartz, topaz), pink (rose quartz), dark brown (smoky topaz), purple (amethyst), etc.; white powder; crystals have a smoothly curving fracture while massive material breaks unevenly; often occurs in six-sided crystals with pointed terminations, also massive, coarsely to finely granular, and as sand; an extremely common mineral which is used for many purposes including, in the case of the transparent, attractively colored varieties, semiprecious gems.

Rhodochrosite (Manganese Spar)

Carbonate of manganese. Easily scratched with a knife; glassy to stony luster; red when transparent and light to deep pink when translucent, but tarnishes black; white powder; splits smoothly and easily in three directions at angles of 73 degrees and 107 degrees with each other; dissolves in hot concentrated muriatic acid with the emission of bubbles; occurs in cleavable or granular masses and in crystals that may have curved faces; a fairly common gangue mineral and is sometimes mined as an ore of manganese of which it contains 41.5 per cent when pure.

Rhodonite (Manganese Spar)

Silicate of manganese. Too hard to be scratched with a knife; stony to glassy luster; red to rose pink, but tarnishes black; white powder; splits in two directions at nearly right angles, but not smoothly like pink orthoclase which it resembles; occurs in cleavable or granular masses or in crystals that are usually rough with rounded edges; used as an ornamental stone and to make violet glass or glaze.

Ruby

Red, gem-quality corundum (which see).

Sapphire

Blue, gem-quality corundum (which see). The term sapphire is also applied by gem miners and cutters to transparent corundum of any color. Some color term is usually prefixed, such as pink sapphire, golden sapphire, purple sapphire, etc.

Siderite (Spathic Iron or Brown Spar)

Carbonate of iron. Easily scratched with a knife; somewhat glassy to dull luster; brown to gray, but tarnishes reddish brown to brownish black; white to yellowish powder; splits easily and smoothly in three directions that make angles of 73 degrees and 107 degrees with each other; dissolves in hot concentrated muriatic acid with the emission of bubbles; occurs in cleavable or coarsely to finely granular masses, earthy masses, and as crystals which often have curved faces; an unimportant ore of iron and sometimes occurs as a gangue mineral.

Silver

Native silver with some gold and other impurities. Easily cut with a knife; metallic luster; silver-white, but easily tarnishes yellowish to brownish or black; may be pounded into thin sheets without breaking; half again as heavy as iron; occurs in masses, grains, scales, and crystals that are usually distorted and connected so as to form fern-like or tree-like aggregates.

Smaltite

Arsenide of cobalt and nickel. Like cobaltite (which see), from which it can be distinguished only by blowpipe or chemical tests; the chief ore of cobalt and often carries silver.

Smithsonite

Carbonate of zinc. Glassy or resinous varieties may be scratched with a knife blade if considerable pressure is exerted, but the earthy variety may seem very soft; glassy or resinous to dull luster; white when pure, but usually gray, brown, yellow, or greenish; white powder; two-thirds heavier than quartz or limestone; dissolves in cold concentrated, but not cold dilute muriatic acid with the emission of bubbles; occurs in crusts with a radiating texture and rounded projections on the surface, as stalactites, and in earthy, often cavernous masses (dry bone); a valuable ore of zinc that is found only in the oxidized zone and contains 52 per cent of that metal when pure.

Sphalerite (Zinc Blende, Jack, Rosin Zinc, Etc.)

Sulphide of zinc. Soft enough to be scratched with a knife; usually resinous luster, but black material is sometimes nearly metallic; commonly brown, but may be yellow, black, red, green, or nearly white; yellow, brown, or almost white powder, always lighter in color than the mineral; splits easily and smoothly in six directions at angles of 60, 90, and 120 degrees; half again as heavy as quartz; occurs in cleavable, coarsely to very finely granular masses that sometimes have spheroidal projections on the surface, occasionally in stalactites, in disseminated grains, and in crystals; often associated with galena; the principal ore of zinc of which metal it contains 67 per cent when pure, and, from some occurrences, most of the cadmium of commerce is obtained.

Spodumene

Silicate of lithium and aluminum. Too hard to be scratched with a knife; stony or dull to glassy luster; usually grayish white, but may be green (hiddenite), pink (kunzite), or colorless; white powder; usually splits or parts easily into plates and often breaks rather roughly in two other directions at 90 degrees with each other and about 45 degrees with the platy parting; occurs in square crystals, sometimes very large, and in cleavable masses that look much like orthoclase; hiddenite and kunzite are valuable gems.

Stephanite (Brittle Silver)

Sulphantimonite of silver. Sometimes soft enough to be scratched with a finger nail; metallic luster; dark gray; dark gray powder; very brittle; uneven fracture; occurs in coarsely to very finely granular masses, disseminated particles, and flat, six-sided crystals; a rare ore of silver that contains 68.5 per cent of that metal when pure.

Stibnite

Sulphide of antimony. May be scratched with a finger nail; bright metallic luster; light gray, but tarnishes iridescent or black; lead-gray powder; breaks easily into flat strips that are marked or bent perpendicular to their longest dimension; occurs in masses with a bladed texture, in deeply grooved crystals, and, rarely, granular; sometimes carries gold and silver and is the chief source of antimony of which it contains 71.8 per cent when pure.

Sulphur (Brimstone)

Native sulphur, sometimes containing selenium and other impurities. May often be scratched with a finger nail; resinous to dull luster; various shades of yellow, orange, brown, or gray, and sometimes nearly white; white or yellowish powder; very brittle; melts in a match flame and yields the stifling odor of sulphur dioxide; occurs in crystals, more or less porous and sometimes earthy masses, as crusts, and as a powder; enormous quantities are used for many different purposes.

Talc (Talcum)

Hydrous silicate of magnesium. Very easily scratched with a finger nail when pure; waxy to dull luster; white, light to dark green, gray, brown, or reddish; white powder; crystalline material breaks easily and smoothly into thin plates that are rather flexible, but not elastic, and may be cut like paper; feels greasy or soapy; occurs in platy or coarsely to finely granular masses; has many industrial uses.

Tantalite

Columbate of iron and manganese. Cannot be distinguished with certainty from columbite (which see) without chemical tests, although the specific gravity of tantalite is usually somewhat higher.

Tennantite (Gray Copper)

Sulpharsenite of copper. May be scratched with a knife; metallic luster; steel gray, sometimes with a slight reddish or brownish tint, powder usually slightly brownish or reddish, but sometimes gray; slightly rough fracture; very brittle; occurs in coarsely to very finely granular masses, and very rarely in crystals; an ore of copper that is often rich in silver and may contain zinc, mercury, lead, tin, or cobalt.

Tenorite (Melaconite or Black Oxide of Copper)

Oxide of copper. May be scratched with a knife, and, often, with a finger nail; metallic when in scales, but otherwise dull luster; iron-gray when in scales, but otherwise black; black powder; scales are slightly flexible; may soil the fingers; occurs in earthy to scaly masses and in tiny metallic lustered scales; an alteration product of other copper minerals with which it is sometimes associated in the oxidized zone.

Tetrahedrite (Gray Copper)

Sulphantimonite of copper. Like tennantite (which see), but is always dark gray in color and the powder is gray.

Titanite (Sphene)

Titano-silicate of calcium. May be scratched with a knife if great pressure is applied; glassy to resinous luster; usually brown to black, but sometimes yellow, and rarely green or pink; powder usually white; often splits or parts smoothly in one or two directions that make angles of about 66½ degrees with each other; usually occurs in platy masses or as flat crystals with wedge-like edges; the transparent, colored varieties are sometimes used as gems although too soft to wear well.

Topaz (Precious Topaz)

Silicate of aluminum with fluorine. Much too hard to be scratched with a knife and easily scratches quartz; glassy luster; usually colorless or white after exposure to light, but often yellow, brown, or a light tint of blue when first mined, and some specimens retain these tints permanently; white powder; breaks easily and smoothly in one direction (perpendicular to the longest dimension of crystals); somewhat heavier than quartz; occurs in cleavable masses, very complex crystals, water-worn pebbles, and, rarely, granular; permanently tinted, transparent material is used as gems.

Tourmaline (Schorl)

Very complex silicate of many bases, with boron. As hard as quartz or a little harder; glassy luster; usually black, but sometimes brown, red, green, or blue (when not black or brown, the color of the interior of a crystal may differ from that of the exterior or the two ends may be of a different color); white powder; smoothly curving to uneven fracture; occurs usually in schist or pegmatite in elongated crystals that have spherical triangular sections (triangles with sides curved outward); the transparent colored varieties are valuable semiprecious gems.

Vanadinite

Vanadate of lead with chlorine. Easily scratched with a knife; resinous luster; red, reddish brown, yellowish brown, or yellow; white to yellowish powder; nearly as heavy as iron; usually occurs in crystals that are six-sided prisms; usually associated with wulfenite (which see) and is confined to the oxidized zone; an important source of vanadium and contains 10.8 per cent of that element when pure.

Wolframite (Tungsten)

Tungstate of iron and manganese. May be scratched with a knife if extreme pressure is applied; nearly a metallic luster; black, dark brown, or dark gray; dark brown to black powder; splits easily and smoothly in one direction; about as heavy as iron; usually occurs in bladed masses or coarsely to finely granular, but is sometimes found in rather flat crystals; an important ore of tungsten that contains 60.7 per cent of that metal when pure.

Wulfenite

Molybdate of lead. Easily scratched with a knife; resinous luster; usually yellow, orange, or red, but sometimes brown or light gray; white powder; very brittle; nearly twice as heavy as quartz; usually occurs in square, flat crystals with beveled edges or in platy masses; usually associated with vanadinite (which see), and is confined to the oxidized zone; a relatively unimportant ore of molybdenum.

Wurtzite

Sulphide of zinc. Very much like sphalerite (which see), but it crystallizes differently, lacks the cleavage of sphalerite, and is apt to occur in fibrous or columnar masses.

APPENDIX B

ROCKS

Five classes of rocks are generally recognized, namely: igneous, clastic, chemical precipitates, organic, and metamorphic. Each class may be briefly defined and described as follows:

IGNEOUS ROCKS

Igneous rocks are formed by the solidification of once molten earth material—magma. Three sub-divisions of igneous rocks may be recognized, namely: plutonic, minor intrusive, and extrusive. Most igneous rocks are very hard when fresh.

Plutonic Rocks

Such igneous rocks have usually cooled slowly far below the surface where the pressure is very high. They ordinarily occur in masses of great size, although relatively narrow dikes of some plutonic species, such as pegmatite, are common in some localities. Plutonic rocks are compact, composed of interlocked grains large enough to be seen with the unaided eyes, which consist of two or more ingredients each of which may be readily recognized by a mineralogist, and, with very rare exceptions, the more plentiful ingredients do not occur in well-formed crystals.

Minor Intrusive Rocks

Such igneous rocks are formed from magma that has risen toward the surface of the earth through cracks (forming dikes) or has spread between layers of earth materials (forming sills, etc.). Part, at least, of the ingredients are forced to crystallize with relative rapidity when the magma comes in contact with cool earth material (resulting in fineness of grain) and, although the pressure on the solidifying magma averages less than on plutonic magmas, it is still relatively great, so the resulting rock is compact. All of the ingredients of some minor intrusive rocks are so small that none of them may be identified or even seen with the unaided eyes, but, typically, these rocks show well-formed crystals of one or two minerals imbedded in a finer grained groundmass.

Extrusive Rocks

Such igneous rocks have solidified on or relatively close to the surface of the earth and occur typically in surface flows, volcanic necks, and dikes. Otherwise, much that was said about minor intrusive rocks applies to extrusive rocks, but the latter are frequently porous, are more commonly composed of lava glass (obsidian), and they are sometimes banded (show flow texture).

CLASTIC ROCKS

Clastic rocks are composed of fragments of other rocks, produced by weathering or mechanical disintegration. The fragments may be several inches or feet in diameter (conglomerates and breccias), much smaller,

94

but visible to the unaided eyes (arkose, grit, and sandstone), or microscopic (shale, clay, and some limestone). The fragments may be transported by wind and water and are usually stratified (laid down in distinct layers), especially if finally deposited in water, when they are called sediments or sedimentary rocks. The fragments are often eventually cemented together more or less firmly, but all degrees of hardness are found in clastic rocks. Such rocks may contain shells or other remains of organisms (fossils).

CHEMICAL PRECIPITATE ROCKS

Chemical precipitate rocks are composed of material deposited by precipitation from water solutions, usually as the result of evaporation. The precipitate rocks are commonly interbedded with sediments that were washed into the evaporating body of water. Rock salt and gypsum are illustrations of this type of rock.

ORGANIC ROCKS

Organic rocks are composed of (1) material secreted or deposited by animals or plants, or (2) made up of animal or plant remains. Illustrations are some limestone and coal.

METAMORPHIC ROCKS

Metamorphic rocks are made up of other rocks that have been changed in appearance or composition, or both, by pressure, heat, or solutions that have percolated through them. Two types are recognized, namely: regional or dynamometamorphic rocks and contact or thermometamorphic rocks.

Regional or Dynamometamorphic Rocks

Such rocks are composed of earth material that has been deeply buried and, therefore, subjected to enormous pressure and some heat. Such rocks are often banded, hard, and composed of interlocked grains visible to the unaided eyes. They sometimes contain well-formed crystals. Illustrations are mica schist, gneiss, slate, quartzite, and some marble.

Contact or Thermometamorphic Rocks

Such rocks are composed of material that has been changed and often baked by the heat of intrusions of molten magma and by the chemically active solutions expelled by such magma as they cool. Illustrations are garnet and epidote rocks.

DESCRIPTION OF ROCK SPECIES MENTIONED IN THIS BOOK

Alaskite

A granular plutonic igneous rock composed almost entirely of the relatively light colored minerals, quartz and feldspar. Like granite (which see), but lacking dark colored constituents.

Andesite

An extrusive igneous rock that usually contains glassy, light colored, plagioclase feldspar crystals in a darker colored, fine-grained groundmass. Does not contain visible quartz, but may contain black hornblende or black augite pyroxene crystals. The groundmass may be porous and the rock then has a very rough fracture.

Basalt (Malpais)

An extrusive igneous rock that often contains small, black crystals of pyroxene or dark green or brown crystals of olivine in a somewhat lighter colored groundmass in which may be imbedded, however, numerous small, very slender, glassy plagioclase feldspar crystals. Usually very dark colored and relatively heavy and frequently decidedly porous.

Breccia

A clastic rock that is composed of angular broken fragments of other rocks, more or less firmly cemented together. The fragments are often large, and must exceed about ⅛ inch in diameter.

Conglomerate

A clastic rock that is composed of large, rounded fragments of other rocks more or less firmly cemented together; otherwise like a breccia (which see).

Diabase

A minor intrusive igneous rock that shows numerous slender, dull or stony lustered, usually white, plagioclase feldspar crystals, pointing in all directions, imbedded in a dark colored (often black on fresh surfaces) groundmass that is composed of pyroxene.

Diorite

A plutonic igneous rock like granite (which see) in texture, but it contains no visible quartz and is usually predominantly composed of white plagioclase feldspar and black hornblende.

Gabbro

A plutonic igneous rock like granite (which see) in texture, but it contains no quartz and is usually predominantly composed of dark colored pyroxene and lesser amounts of white or light green plagioclase feldspar.

Gneiss

A regional metamorphic rock usually associated with other schists and much like granite (which see) in composition. In fact small specimens cannot always be distinguished from granite, but the rock is plainly banded when seen in the field, and white mica (muscovite) is a very common constituent.

Granite

A plutonic igneous rock that consists essentially of pinkish to white orthoclase feldspar and quartz, but mica, especially black mica (biotite), is a common constituent and other minerals may be present in subordinate amounts.

Limestone

A rock composed essentially of carbonate of lime which dissolves with the emission of bubbles when a drop of dilute or concentrated muriatic acid is placed upon it. Depending upon its origin it may be either a clastic or an organic rock or a chemical precipitate. Most limestones are finely granular and light colored. They are often flint-like in appearance when freshly broken. They may contain fossils which are usually composed of quartz.

Marble

A regional or contact metamorphic rock that is formed from limestone and has the same composition as limestone, but it is rather coarsely granular so that the individual grains, which are usually glassy, and have the perfect cleavage of calcite (which see), are visible to the unaided eyes.

Monzonite

A plutonic igneous rock like diorite (which see), but it contains both orthoclase (often pinkish in color) and plagioclase (often white or greenish in color) feldspar.

Pegmatite

A plutonic igneous rock that occurs in dikes and is much like granite (which see) in texture and composition, but the individual grains or crystals are very large (often several inches long, or larger), and white mica (muscovite) is a much commoner constituent than black mica (biotite). May contain beryl, topaz, tourmaline, and even rarer minerals.

Peridotite

A basic plutonic rock that is usually dark colored and relatively heavy. It contains no feldspar or quartz. Dark brown or green pyroxene (hypersthene, enstatite, or diallage) and olivine are the most plentiful ingredients, but it may also contain magnetite, chromite, and pyrrhotite.

Phonolite

An extrusive igneous rock that is described in Chapter V.

Porphyry

A minor intrusive igneous rock that, typically, shows well-formed crystals of light colored, stony lustered orthoclase feldspar, and, more rarely, quartz in a dense, fine-grained groundmass.

Quartzite

A regional metamorphic rock formed from sandstone. The openings between the grains in the sandstone have been filled with quartz and the resulting rock is very dense.

Sandstone

A clastic rock that is composed of grains of other substances, usually largely or almost entirely quartz, which are more or less firmly cemented by silica, calcium carbonate, iron oxide, or some other substance. The individual grains are visible to unaided eyes and are of the size of coarse granulated sugar, or smaller.

Schist

A regional metamorphic rock that has, typically, a banded (schistose) texture and often breaks readily along the bands. There are many varieties, such as mica schist, tremolite schist, etc., each being usually named by prefixing the name of the most prominent mineral or minerals.

Serpentine

A rock formed by the alteration of very basic igneous rocks like peridotite (which see). It is usually soft enough to be easily scratched with a knife, has a somewhat waxy or greasy luster, feels smooth, breaks with a smoothly curving fracture, is most often some shade of green (commonly dark) in color, and light may frequently be seen through thin edges.

Shale

A clastic rock made of layers (often as thin as cardboard, or thinner) of particles, too small to be visible to unaided eyes, of various hydrous aluminum silicates (of which kaolinite is one), quartz, and other impurities. It is usually soft, smells like clay, especially when moistened, and breaks along the layers. The color is usually brown or gray.

Slate

A regional metamorphic rock like shale in composition but much harder. It breaks into strong, thin sheets perpendicular to the pressure that produced it and the color is most frequently black.

Syenite

A plutonic igneous rock like granite (which see), but it contains no visible quartz and black hornblende is more often present than is black mica.

Rhyolite

An extrusive igneous rock that is usually light colored and relatively light in weight. It often contains glassy, colorless or white orthoclase feldspar crystals and, less frequently, quartz crystals imbedded in a fine grained groundmass that often feels rough. It sometimes shows flow texture (is banded) and may be glassy (obsidian). When it is very finely porous and contains no grains of minerals visible to unaided eyes, it is called pumice.

Trachyte

Exactly like rhyolite (which see), but no quartz, or practically none, is visible even in thin sections under the compound miscroscope. It cannot usually be distinguished from rhyolite in the field.

Tuff

Composed mostly of the fine material (particles of volcanic glass) called volcanic ash that is thrown into the air during volcanic eruptions. It is usually white or light tinted, porous, light in weight, and soft. It feels rough and commonly contains angular fragments of extrusive rocks.

Volcanic Breccia

A breccia (which see) that contains numerous angular fragments of extrusive rocks.

SELECTED BIBLIOGRAPHY

Minerals

Minerals and How to Study Them, Edw. S. Dana, John Wiley and Sons, Inc., 440 4th Ave., New York, $2.00.
(Intended to assist the beginner to recognize the commoner minerals and to serve as a basis for further study.)

Pocket Handbook of Minerals, G. Montague Butler, John Wiley and Sons, Inc., 440 4th Ave., New York, $3.00.
(A pocket reference work in which emphasis is placed upon characteristic physical features.)

Rocks

Rocks and Rock Minerals, L. V. Pirrson and Adolph Knopf, John Wiley and Sons, 440 4th Ave., New York, $3.50.
(A very good introduction to a rather difficult subject.)

Prospecting

Handbook for Prospectors, M. W. von Bernewitz, McGraw-Hill Book Co., Inc., 370 7th Ave., New York, $3.00.
(Contains much useful information for prospectors.)

Mining Geology

Geology Applied to Mining, J. E. Spurr, McGraw-Hill Book Company, 370 7th Ave., New York, $3.00.
(A very useful book for miners who wish to increase their knowledge of the subject covered by it.)

Ore Deposits

Mineral Deposits, Waldemar Lindgren, McGraw-Hill Book Company, 370 7th Ave., New York, $7.00.
(A voluminous and authoritative book covering all types of ore deposits.)

The Principles of Economic Geology, W. H. Emmons, McGraw-Hill Book Company, 370 7th Ave., New York, $5.00.
(A somewhat briefer treatment of the subject.)

Miscellaneous

Leached Outcrops as Guides to Copper Ore, Augustus Locke, The Williams and Wilkins Co., Baltimore, Md., $5.00.
(The first book to appear on this important subject.)

Non Metallic Minerals, Occurrence-Preparation-Utilization, B. Ladoo, McGraw-Hill Book Company, 370 7th Ave., New York, $6.00.
(A book that will prove very useful to persons interested in nonmetallics.)

The Marketing of Metals and Miners, J. E. Spurr and F. E. Wormser, McGraw-Hill Book Company, 370 7th Ave., New York, $6.00.
(Every mining man will-find valuable information in this book.)

GOLD RUSH BOOKS

OREGON, USA

www.GoldMiningBooks.com

Books On Mining

Visit: www.goldminingbooks.com to order your copies or ask your favorite book seller to offer them.

Mining Books by Kerby Jackson

<u>Gold Dust: Stories From Oregon's Mining Years</u> - Oregon mining historian and prospector, Kerby Jackson, brings you a treasure trove of seventeen stories on Southern Oregon's rich history of gold prospecting, the prospectors and their discoveries, and the breathtaking areas they settled in and made homes. 5" X 8", 98 ppgs. Retail Price: $11.99

<u>The Golden Trail: More Stories From Oregon's Mining Years</u> - In his follow-up to "Gold Dust: Stories of Oregon's Mining Years", this time around, Jackson brings us twelve tales from Oregon's Gold Rush, including the story about the first gold strike on Canyon Creek in Grant County, about the old timers who found gold by the pail full at the Victor Mine near Galice, how Iradel Bray discovered a rich ledge of gold on the Coquille River during the height of the Rogue River War, a tale of two elderly miners on the hunt for a lost mine in the Cascade Mountains, details about the discovery of the famous Armstrong Nugget and others. 5" X 8", 70 ppgs. Retail Price: $10.99

Oregon Mining Books

<u>Geology and Mineral Resources of Josephine County, Oregon</u> - Unavailable since the 1970's, this important publication was originally compiled by the Oregon Department of Geology and Mineral Industries and includes important details on the economic geology and mineral resources of this important mining area in South Western Oregon. Included are notes on the history, geology and development of important mines, as well as insights into the mining of gold, copper, nickel, limestone, chromium and other minerals found in large quantities in Josephine County, Oregon. 8.5" X 11", 54 ppgs. Retail Price: $9.99

<u>Mines and Prospects of the Mount Reuben Mining District</u> - Unavailable since 1947, this important publication was originally compiled by geologist Elton Youngberg of the Oregon Department of Geology and Mineral Industries and includes detailed descriptions, histories and the geology of the Mount Reuben Mining District in Josephine County, Oregon. Included are notes on the history, geology, development and assay statistics, as well as underground maps of all the major mines and prospects in the vicinity of this much neglected mining district. 8.5" X 11", 48 ppgs. Retail Price: $9.99

<u>The Granite Mining District</u> - Notes on the history, geology and development of important mines in the well known Granite Mining District which is located in Grant County, Oregon. Some of the mines discussed include the Ajax, Blue Ribbon, Buffalo, Continental, Cougar-Independence, Magnolia, New York, Standard and the Tillicum. Also included are many rare maps pertaining to the mines in the area. 8.5" X 11", 48 ppgs. Retail Price: $9.99

<u>Ore Deposits of the Takilma and Waldo Mining Districts of Josephine County, Oregon</u> - The Waldo and Takilma mining districts are most notable for the fact that the earliest large scale mining of placer gold and copper in Oregon took place in these two areas. Included are details about some of the earliest large gold mines in the state such as the Llano de Oro, High Gravel, Cameron, Platerica, Deep Gravel and others, as well as copper mines such as the famous Queen of Bronze mine, the Waldo, Lily and Cowboy mines. This volume also includes six maps and 20 original illustrations. 8.5" X 11", 74 ppgs. Retail Price: $9.99

<u>Metal Mines of Douglas, Coos and Curry Counties, Oregon</u> - Oregon mining historian Kerby Jackson introduces us to a classic work on Oregon's mining history in this important re-issue of Bulletin 14C Volume 1, otherwise known as the Douglas, Coos & Curry Counties, Oregon Metal Mines Handbook. Unavailable since 1940, this important publication was originally compiled by the Oregon Department of Geology and Mineral Industries includes detailed descriptions, histories and the geology of over 250 metallic mineral mines and prospects in this rugged area of South West Oregon. 8.5" X 11", 158 ppgs. Retail Price: $19.99

Metal Mines of Jackson County, Oregon - Unavailable since 1943, this important publication was originally compiled by the Oregon Department of Geology and Mineral Industries includes detailed descriptions, histories and the geology of over 450 metallic mineral mines and prospects in Jackson County, Oregon. Included are such famous gold mining areas as Gold Hill, Jacksonville, Sterling and the Upper Applegate. **8.5" X 11", 220 ppgs. Retail Price: $24.99**

Metal Mines of Josephine County, Oregon - Oregon mining historian Kerby Jackson introduces us to a classic work on Oregon's mining history in this important re-issue of Bulletin 14C, otherwise known as the Josephine County, Oregon Metal Mines Handbook. Unavailable since 1952, this important publication was originally compiled by the Oregon Department of Geology and Mineral Industries includes detailed descriptions, histories and the geology of over 500 metallic mineral mines and prospects in Josephine County, Oregon. **8.5" X 11", 250 ppgs. Retail Price: $24.99**

Metal Mines of North East Oregon - Oregon mining historian Kerby Jackson introduces us to a classic work on Oregon's mining history in this important re-issue of Bulletin 14A and 14B, otherwise known as the North East Oregon Metal Mines Handbook. Unavailable since 1941, this important publication was originally compiled by the Oregon Department of Geology and Mineral Industries and includes detailed descriptions, histories and the geology of over 750 metallic mineral mines and prospects in North Eastern Oregon. **8.5" X 11", 310 ppgs. Retail Price: $29.99**

Metal Mines of North West Oregon - Oregon mining historian Kerby Jackson introduces us to a classic work on Oregon's mining history in this important re-issue of Bulletin 14D, otherwise known as the North West Oregon Metal Mines Handbook. Unavailable since 1951, this important publication was originally compiled by the Oregon Department of Geology and Mineral Industries and includes detailed descriptions, histories and the geology of over 250 metallic mineral mines and prospects in North Western Oregon. **8.5" X 11", 182 ppgs. Retail Price: $19.99**

Mines and Prospects of Oregon - Mining historian Kerby Jackson introduces us to a classic mining work by the Oregon Bureau of Mines in this important re-issue of The Handbook of Mines and Prospects of Oregon. Unavailable since 1916, this publication includes important insights into hundreds of gold, silver, copper, coal, limestone and other mines that operated in the State of Oregon around the turn of the 19th Century. Included are not only geological details on early mines throughout Oregon, but also insights into their history, production, locations and in some cases, also included are rare maps of their underground workings. **8.5" X 11", 314 ppgs. Retail Price: $24.99**

Lode Gold of the Klamath Mountains of Northern California and South West Oregon
(See California Mining Books)

Mineral Resources of South West Oregon - Unavailable since 1914, this publication includes important insights into dozens of mines that once operated in South West Oregon, including the famous gold fields of Josephine and Jackson Counties, as well as the Coal Mines of Coos County. Included are not only geological details on early mines throughout South West Oregon, but also insights into their history, production and locations. **8.5" X 11", 154 ppgs. Retail Price: $11.99**

Chromite Mining in The Klamath Mountains of California and Oregon
(See California Mining Books)

Southern Oregon Mineral Wealth - Unavailable since 1904, this rare publication provides a unique snapshot into the mines that were operating in the area at the time. Included are not only geological details on early mines throughout South West Oregon, but also insights into their history, production and locations. Some of the mining areas include Grave Creek, Greenback, Wolf Creek, Jump Off Joe Creek, Granite Hill, Galice, Mount Reuben, Gold Hill, Galls Creek, Kane Creek, Sardine Creek, Birdseye Creek, Evans Creek, Foots Creek, Jacksonville, Ashland, the Applegate River, Waldo, Kerby and the Illinois River, Althouse and Sucker Creek, as well as insights into local copper mining and other topics. **8.5" X 11", 64 ppgs. Retail Price: $8.99**

Geology and Ore Deposits of the Takilma and Waldo Mining Districts - Unavailable since the 1933, this publication was originally compiled by the United States Geological Survey and includes details on gold and copper mining in the Takilma and Waldo Districts of Josephine County, Oregon. The Waldo and Takilma mining districts are most notable for the fact that the earliest large scale mining of placer gold and copper in Oregon took place in these two areas. Included in this report are details about some of the earliest large gold mines in the state such as the Llano de Oro, High Gravel, Cameron, Platerica, Deep Gravel and others, as well as copper mines such as the famous Queen of Bronze mine, the Waldo, Lily and Cowboy mines. In addition to geological examinations, insights are also provided into the production, day to day operations and early histories of these mines, as well as calculations of known mineral reserves in the area. This volume also includes six maps and 20 original illustrations. **8.5" X 11", 74 ppgs. Retail Price: $9.99**

Gold Mines of Oregon - Oregon mining historian Kerby Jackson introduces us to a classic work on Oregon's mining history in this important re-issue of Bulletin 61, otherwise known as "Gold and Silver In Oregon". Unavailable since 1968, this important publication was originally compiled by geologists Howard C. Brooks and Len Ramp of the Oregon Department of Geology and Mineral Industries and includes detailed descriptions, histories and the geology of over 450 gold mines Oregon. Included are notes on the history, geology and gold production statistics of all the major mining areas in Oregon including the Klamath Mountains, the Blue Mountains and the North Cascades. While gold is where you find it, as every miner knows, the path to success is to prospect for gold where it was previously found. **8.5" X 11", 344 ppgs. Retail Price: $24.99**

Mines and Mineral Resources of Curry County Oregon - Originally published in 1916, this important publication on Oregon Mining has not been available for nearly a century. Included are rare insights into the history, production and locations of dozens of gold mines in Curry County, Oregon, as well as detailed information on important Oregon mining districts in that area such as those at Agness, Bald Face Creek, Mule Creek, Boulder Creek, China Diggings, Collier Creek, Elk River, Gold Beach, Rock Creek, Sixes River and elsewhere. Particular attention is especially paid to the famous beach gold deposits of this portion of the Oregon Coast. **8.5" X 11", 140 ppgs. Retail Price: $11.99**

Chromite Mining in South West Oregon - Originally published in 1961, this important publication on Oregon Mining has not been available for nearly a century. Included are rare insights into the history, production and locations of nearly 300 chromite mines in South Western Oregon. **8.5" X 11", 184 ppgs. Retail Price: $14.99**

Mineral Resources of Douglas County Oregon - Originally published in 1972, this important publication on Oregon Mining has not been available for nearly forty years. Included are rare insights into the geology, history, production and locations of numerous gold mines and other mining properties in Douglas County, Oregon. **8.5" X 11", 124 ppgs. Retail Price: $11.99**

Mineral Resources of Coos County Oregon - Originally published in 1972, this important publication on Oregon Mining has not been available for nearly forty years. Included are rare insights into the geology, history, production and locations of numerous gold mines and other mining properties in Coos County, Oregon. **8.5" X 11", 100 ppgs. Retail Price: $11.99**

Mineral Resources of Lane County Oregon - Originally published in 1938, this important publication on Oregon Mining has not been available for nearly seventy five years. Included are extremely rare insights into the geology and mines of Lane County, Oregon, in particular in the Bohemia, Blue River, Oakridge, Black Butte and Winberry Mining Districts. **8.5" X 11", 82 ppgs. Retail Price: $9.99**

Mineral Resources of the Upper Chetco River of Oregon: Including the Kalmiopsis Wilderness - Originally published in 1975, this important publication on Oregon Mining has not been available for nearly forty years. Withdrawn under the 1872 Mining Act since 1984, real insight into the minerals resources and mines of the Upper Chetco River has long been unavailable due to the remoteness of the area. Despite this, the decades of battle between property owners and environmental extremists over the last private mining inholding in the area has continued to pique the interest of those interested in mining and other forms of natural resource use. Gold mining began in the area in the 1850's and has a rich history in this geographic area, even if the facts surrounding it are little known. Included are twenty two rare photographs, as well as insights into the Becca and Morning Mine, the Emmly Mine (also known as Emily Camp), the Frazier Mine, the Golden Dream or Higgins Mine, Hustis Mine, Peck Mine and others. **8.5" X 11", 64 ppgs. Retail Price: $8.99**

Gold Dredging in Oregon - Originally published in 1939, this important publication on Oregon Mining has not been available for nearly seventy five years. Included are extremely rare insights into the history and day to day operations of the dragline and bucketline gold dredges that once worked the placer gold fields of South West and North East Oregon in decades gone by. Also included are details into the areas that were worked by gold dredges in Josephine, Jackson, Baker and Grant counties, as well as the economic factors that impacted this mining method. This volume also offers a unique look into the values of river bottom land in relation to both farming and mining, in how farm lands were mined, re-soiled and reclamated after the dredges worked them. Featured are hard to find maps of the gold dredge fields, as well as rare photographs from a bygone era. **8.5" X 11", 86 ppgs. Retail Price: $8.99**

Quick Silver Mining in Oregon - Originally published in 1963, this important publication on Oregon Mining has not been available for over fifty years. This publication includes details into the history and production of Elemental Mercury or Quicksilver in the State of Oregon. **8.5" X 11", 238 ppgs. Retail Price: $15.99**

Mines of the Greenhorn Mining District of Grant County Oregon - Originally published in 1948, this important publication on Oregon Mining has not been available for over sixty five years. In this publication are rare insights into the mines of the famous Greenhorn Mining District of Grant County, Oregon, especially the famous Morning Mine. Also included are details on the Tempest, Tiger, Bi-Metallic, Windsor, Psyche, Big Johnny, Snow Creek, Banzette and Paramount Mines, as well as prospects in the vicinities in the famous mining areas of Mormon Basin, Vinegar Basin and Desolation Creek. Included are hard to find mine maps and dozens of rare photographs from the bygone era of Grant County's rich mining history. **8.5" X 11", 72 ppgs. Retail Price: $9.99**

Geology of the Wallowa Mountains of Oregon: Part I (Volume 1) - Originally published in 1938, this important publication on Oregon Mining has not been available for nearly seventy five years. Included are details on the geology of this unique portion of North Eastern Oregon. This is the first part of a two book series on the area. Accompanying the text are rare photographs and historic maps.**8.5" X 11", 92 ppgs. Retail Price: $9.99**

Geology of the Wallowa Mountains of Oregon: Part II (Volume 2) - Originally published in 1938, this important publication on Oregon Mining has not been available for nearly seventy five years. Included are details on the geology of this unique portion of North Eastern Oregon. This is the first part of a two book series on the area. Accompanying the text are rare photographs and historic maps.**8.5" X 11", 94 ppgs. Retail Price: $9.99**

Field Identification of Minerals For Oregon Prospectors - Originally published in 1940, this important publication on Oregon Mining has not been available for nearly seventy five years. Included in this volume is an easy system for testing and identifying a wide range of minerals that might be found by prospectors, geologists and rockhounds in the State of Oregon, as well as in other locales. Topics include how to put together your own field testing kit and how to conduct rudimentary tests in the field. This volume is written in a clear and concise way to make it useful even for beginners. **8.5" X 11", 158 ppgs. Retail Price: $14.99**

The Bohemia Mining District of Oregon - Originally published in 1900, this important publication on Oregon Mining has not been available for over a century. Included in this volume are important insights into the famous Bohemia Mining District of Oregon, including the histories and locations of important gold mines in the area such as the Ophir Mine, Clarence, Acturas, Peek-a-boo, White Swan, Combination Mine, the Musick Mine, The California, White Ghost, The Mystery, Wall Street, Vesuvius, Story, Lizzie Bullock, Delta, Elsie Dora, Golden Slipper, Broadway, Champion Mine, Knott, Noonday, Helena, White Wings, Riverside and others. Also included are notes on the nearby Blue River Mining District. **8.5" X 11", 58 ppgs. Retail Price: $9.99**

The Gold Fields of Eastern Oregon - Unavailable since 1900, this publication was originally compiled by the Baker City Chamber of Commerce Offering important insights into the gold mining history of Eastern Oregon, "The Gold Fields of Eastern Oregon" sheds a rare light on many of the gold mines that were operating at the turn of the 19th Century in Baker County and Grant County in North Eastern Oregon. Some of the areas featured include the Cable Cove District, Baisely-Elhorn, Granite, Red Boy, Bonanza, Susanville, Sparta, Virtue, Vaughn, Sumpter, Burnt River, Rye Valley and other mining districts. Included is basic information on not only many gold mines that are well known to those interested in Eastern Oregon mining history, but also many mines and prospects which have been mostly lost to the passage of time. Accompanying are numerous rare photos **8.5" X 11", 78 ppgs. Retail Price: $10.99**

Gold Mining in Eastern Oregon - Originally published in 1938, this important publication on Oregon Mining has not been available for over a century. Included in this volume are important insights into the famous mining districts of Eastern Oregon during the late 1930's. Particular attention is given to those gold mines with milling and concentrating facilities in the Greenhorn, Red Boy, Alamo, Bonanza, Granite, Cable Cove, Cracker Creek, Virtue, Keating, Medical Springs, Sanger, Sparta, Chicken Creek, Mormon Basin, Connor Creek, Cornucopia and the Bull Run Mining Districts. Some of the mines featured include the Ben Harrison, North Pole-Columbia, Highland Maxwell, Baisley-Elkhorn, White Swan, Balm Creek, Twin Baby, Gem of Sparta, New Deal, Gleason, Gifford-Johnson, Cornucopia, Record, Bull Run, Orion and others. Of particular interest are the mill flow sheets and descriptions of milling operations of these mines. **8.5" X 11", 68 ppgs. Retail Price: $8.99**

The Gold Belt of the Blue Mountains of Oregon - Originally published in 1901, this important publication on Oregon Mining has not been available for over a century. Included in this volume are rare insights into the gold deposits of the Blue Mountains of North East Oregon, including the history of their early discovery and early production. Extensive details are offered on this important mining area's mineralogy and economic geology, as well as insights into nearby gold placers, silver deposits and copper deposits. Featured are the Elkhorn and Rock Creek mining districts, the Pocahontas district, Auburn and Minersville districts, Sumpter and Cracker Creek, Cable Cove, the Camp Carson district, Granite, Alamo, Greenhorn, Robinsonville, the Upper Burnt River Valley and Bonanza districts, Susanville, Quartzburg, Canyon Creek, Virtue, the Copper Butte district, the North Powder River, Sparta, Eagle Creek, Cornucopia, Pine Creek, Lower Powder River, the Upper Snake River Canyon, Rye Valley, Lower Burnt River Valley, Mormon Basin, the Malheur and Clarks Creek districts, Sutton Creek and others. Of particular interest are important details on numerous gold mines and prospects in these mining districts, including their locations, histories, geology and other important information, as well as information on silver, copper and fire opal deposits. **8.5" X 11", 250 ppgs. Retail Price: $24.99**

Mining in the Cascades Range of Oregon - Originally published in 1938, this important publication on Oregon Mining has not been available for over seventy five years. Included in this volume are rare insights into the gold mines and other types of metal mines in the Cascades Mountain Range of Oregon. Some of the important mining areas covered include the famous Bohemia Mining District, the North Santiam Mining District, Quartzville Mining District, Blue River Mining District, Fall Creek Mining District, Oakridge District, Zinc District, Buzzard-Al Sarena District, Grand Cove, Climax District and Barron Mining District. Of particular interest are important details on over 100 mines and prospects in these mining districts, including their locations, histories, geology and other important information. **8.5″ X 11″, 170 ppgs. Retail Price: $14.99**

Beach Gold Placers of the Oregon Coast - Originally published in 1934, this important publication on Oregon Mining has not been available for over 80 years. Included in this volume are rare insights into the beach gold deposits of the State of Oregon, including their locations, occurance, composition and geology. Of particular interest is information on placer platinum in Oregon's rich beach deposits. Also included are the locations and other information on some famous Oregon beach mines, including the Pioneer, Eagle, Chickamin, Iowa and beach placer mines north of the mouth of the Rogue River. **8.5″ X 11″, 60 ppgs. Retail Price: $8.99**

Mineralogical Composition of the Sands of the Oregon Coast: From Coos Bay to the Columbia – Published in 1945, he text features hard to find information on the composition of the gold bearing black sands of the South West Oregon Coast, offering a unique insight to prospectors in search of Oregon's legendary beach gold. 104 ppgs, $9.99

Manganese Mining in Oregon - First released in 1942 and now out of print, this special reprint edition of "Manganese in Oregon" was originally published by the Oregon Department of Geology and Mineral Industries. The text features hard to find information on the mining of Manganese in Oregon, including details and maps of Oregon manganese mines and prospects. 108 ppgs, 9.99

Medford Oregon As A Mining Center - Written in 1912, this hard to find publication includes valuable insights into the mining history of South West Oregon. This small book contains interesting information on the gold, copper and mining industry in Southern Oregon as it existed just prior to World War One, shedding light on some of the important mines in the area. Included are rare photographs and vintage advertising of the day. 80 ppgs, 9.99

Mineral Resources of Curry County Oregon - First released in 1977 and now out of print, this special reprint edition of "Geology, Mineral Resources and Rock Materials of Curry County, Oregon" was originally published in cooperation of Curry County, Oregon and the Oregon Department of Geology and Mineral Industries. The text features hard to find information on not only the mining of gold and other metals in Curry County, but also aggregate mining in the area. 102 ppgs, 11.99

Origin of the Gold Bearing Black Sands of the Coast of South West Oregon - First released in 1943 and now out of print, this special reprint edition of "The Origin of the Black Sands of the South West Oregon Coast" was originally published by the Oregon Department of Geology and Mineral Industries. The text features hard to find information on the origin of the gold bearing black sands of the South West Oregon Coast, offering a unique insight to prospectors in search of Oregon's legendary beach gold. 52 ppgs, 8.99

South West Oregon Mining - Leading mining historian Kerby Jackson introduces us to six classic small mining publications on the Gold Mining Industry in Southern Oregon. This small book consists of a compilation of USGS J.S. Diller's "Mines of the Riddles Quadrangle", "The Rogue River Valley Coal Fields" and "Mineral Resources of the Grants Pass Quadrangle", the Grants Pass Commercial Club's rare publication "Mining in Josephine County, Oregon" and the USGS publication "The Distribution of Placer Gold in the Sixes River, South West Oregon". Also included is F.W. Libbey's legendary article on the Southern Oregon Mining Industry, "Lest We Forget", which appeared in the publication of the Oregon State Department of Geology and Mineral Industries in the early 1960's. This compilation offers a unique perspective on mining in South West Oregon and includes considerable information on mines in Josephine, Jackson and Coos Counties. 142 ppgs, 14.99

Geology and Mineral Resources of the Gasquet Quadrangle of California-Oregon - First published in 1953, it has been unavailable for over a century and sheds important light on the geological features and mineral resources of this portion of Northern California and Southern Oregon. 80 ppgs, 9.99

Idaho Mining Books

Gold in Idaho - Unavailable since the 1940's, this publication was originally compiled by the Idaho Bureau of Mines and includes details on gold mining in Idaho. Included is not only raw data on gold production in Idaho, but also valuable insight into where gold may be found in Idaho, as well as practical information on the gold bearing rocks and other geological features that will assist those looking for placer and lode gold in the State of Idaho. This volume also includes thirteen gold maps that greatly enhance the practical usability of the information contained in this small book detailing where to find gold in Idaho. **8.5" X 11", 72 ppgs. Retail Price: $9.99**

Geology of the Couer D'Alene Mining District of Idaho - Unavailable since 1961, this publication was originally compiled by the Idaho Bureau of Mines and Geology and includes details on the mining of gold, silver and other minerals in the famous Coeur D'Alene Mining District in Northern Idaho. Included are details on the early history of the Coeur D'Alene Mining District, local tectonic settings, ore deposit features, information on the mineral belts of the Osburn Fault, as well as detailed information on the famous Bunker Hill Mine, the Dayrock Mine, Galena Mine, Lucky Friday Mine and the infamous Sunshine Mine. This volume also includes sixteen hard to find maps. **8.5" X 11", 70 ppgs. Retail Price: $9.99**

The Gold Camps and Silver Cities of Idaho - Originally published in 1963, this important publication on Idaho Mining has not been available for nearly fifty years. Included are rare insights into the history of Idaho's Gold Rush, as well as the mad craze for silver in the Idaho Panhandle. Documented in fine detail are the early mining excitements at Boise Basin, at South Boise, in the Owyhees, at Deadwood, Long Valley, Stanley Basin and Robinson Bar, at Atlanta, on the famous Boise River, Volcano, Little Smokey, Banner, Boise Ridge, Hailey, Leesburg, Lemhi, Pearl, at South Mountain, Shoup and Ulysses, Yellow Jacket and Loon Creek. The story follows with the appearance of Chinese miners at the new mining camps on the Snake River, Black Pine, Yankee Fork, Bay Horse, Clayton, Heath, Seven Devils, Gibbonsville, Vienna and Sawtooth City. Also included are special sections on the Idaho Lead and Silver mines of the late 1800's, as well as the mining discoveries of the early 1900's that paved the way for Idaho's modern mining and mineral industry. Lavishly illustrated with rare historic photos, this volume provides a one of a kind documentary into Idaho's mining history that is sure to be enjoyed by not only modern miners and prospectors who still scour the hills in search of nature's treasures, but also those enjoy history and tromping through overgrown ghost towns and long abandoned mining camps. **8.5" X 11", 186 ppgs. Retail Price: $14.99**

Ore Deposits and Mining in North Western Custer County Idaho - Unavailable since 1913, this important publication was originally published by the Us Department of the Interior and has been unavailable for a century. Included are fine details on the geology, geography, gold placers and gold and silver bearing quartz veins of the mining region of North West Custer County, Idaho. Of particular interest is a rare look at the mines and prospects of the region, including those such as the Ramshorn Mine, SkyLark, Riverview, Excelsior, Beardsley, Pacific, Hoosier, Silver Brick, Forest Rose and dozens of others in the Bay Horse Mining District. Also covered are the mines of the Yankee Fork District such as the Lucky Boy, Badger, Black, Enterprise, Charles Dickens, Morrison, Golden Sunbeam, Montana, Golden Gate and others, as well as those in the Loon Mining District. **8.5" X 11", 126 ppgs. Retail Price: $12.99**

Gold Rush To Idaho - Unavailable since 1963, this important publication was originally published by the Idaho Bureau of Mines and has been unavailable for 50 years. "Gold Rush To Idaho" revisits the earliest years of the discovery of gold in Idaho Territory and introduces us to the conditions that the pioneer gold seekers met when they blazed a trail through the wilderness of Idaho's mountains and discovered the precious yellow metal at Oro Fino and Pierce. Subsequent rushes followed at places like Elk City, Newsome, Clearwater Station, Florence, Warrens and elsewhere. Of particular interest is a rare look at the hardships that the first miners in Idaho met with during their day to day existences and their attempts to bring law and order to their mining camps. **8.5" X 11", 88 ppgs. Retail Price: $9.99**

The Geology and Mines of Northern Idaho and North Western Montana - Unavailable since 1909, this important publication was originally published by the Us Department of the Interior and has been unavailable for a century. Included are fine details on the geology and geography of the mining regions of Northern Idaho and North Western Montana. Of particular interest is a rare look at the mines and prospects of the region, including those in the Pine Creek Mining District, Lake Pend Oreille district, Troy Mining District, Sylvanite District, Cabinet Mining District, Prospect Mining District and the Missoula Valley. Some of the mines featured include the Iron Mountain, Silver Butte, Snowshoe, Grouse Mountain Mine and others. **8.5" X 11", 142 ppgs. Retail Price: $12.99**

Mining in the Alturas Quadrangle of Blaine County Idaho - Unavailable since 1922, this important publication was originally published by the Idaho Bureau of Mines and has been unavailable for ninety years. Topics include the geology, rock formations and the formation of ore deposits in this important mining area of Idaho. Of particular focus is information on the local geology, quartz veins and ore deposits of this portion of Idaho. Included are hard to find details, including the descriptions and locations of numerous gold and silver mines in the area including the Silver King, Pilgrim, Columbia, Lone Jack, Sunbeam, Pride of the West, Lucky Boy, Scotia, Atlanta, Beaver-Bidwell and others mines and prospects. **8.5" X 11", 56 ppgs. Retail Price: $8.99**

Mining in Lemhi County Idaho - Originally published in 1913, this important book on Idaho Mining has not been available to miners for over a century. Included are rare insights into hundreds of gold, silver, copper and other mines in this famous Idaho mining area. Details include the locations, geology, history, production and other facts of the mines of this region, not only gold and silver hardrock mines, but also gold placer mines, lead-silver deposits, copper mines, cobalt-nickel deposits, tungsten and tin mines . It is lavishly illustrated with hard to find photos of the period and rare mining maps. Some of the vicinities featured include the Nicholia Mining District, Spring Mountain District, Texas District, Blue Wing District, Junction District, McDevitt District, Pratt Creek, Eldorado District, Kirtley Creek, Carmen Creek, Gibbonsville, Indian Creek, Mineral Hill District, Mackinaw, Eureka District, Blackbird District, YellowJacket District, Gravel Range District, Junction District, Parker Mountain and other mining districts. 8.5" X 11", 226 ppgs. Retail Price: $19.99

Mining in Shoshone County Idaho - First published in 1923, it has been unavailable for over a century and sheds important light on the mining history of Shoshone County, Idaho. Some of the topics include the history of mining in Shoshone County, a look at the local geology and ore characteristics of lead-silver deposits, zinc deposits, copper, antimony, gold and other minerals. Also included are insights into the history, production, characteristics and locations of numerous mines in the area. 198 ppgs, 15.99

Utah Mining Books

Fluorite in Utah - Unavailable since 1954, this publication was originally compiled by the USGS, State of Utah and U.S. Atomic Energy Commission and details the mining of fluorspar, also known as fluorite in the State of Utah. Included are details on the geology and history of fluorspar (fluorite) mining in Utah, including details on where this unique gem mineral may be found in the State of Utah. 8.5" X 11", 60 ppgs. Retail Price: $8.99

The Gold Hill Mining District of Utah - First published in 1935, it has been unavailable since those days and sheds important light on the mines, history and geology of Utah's Gold Hill Mining District. Included are rare insights into this important mining area, including the locations, histories and details of numerous mines. This volume is well illustrated with geological diagrams, as well as hard to find maps of some of the most important mines in this district. 202 ppgs., 19.99

The Mines, Miners and Minerals of Utah - First published in 1896, it has been unavailable since those days and sheds important light on the early mines and miners of Pioneer Utah, as well as the minerals which they won from the earth by laborious hard physical labor and sheer determination. Included are rare insights into the early mining history of Utah, as well details on hundreds of gold, silver and copper mines. 376 ppgs., 24.99

California Mining Books

The Tertiary Gravels of the Sierra Nevada of California - Mining historian Kerby Jackson introduces us to a classic mining work by Waldemar Lindgren in this important re-issue of The Tertiary Gravels of the Sierra Nevada of California. Unavailable since 1911, this publication includes details on the gold bearing ancient river channels of the famous Sierra Nevada region of California. 8.5" X 11", 282 ppgs. Retail Price: $19.99

The Mother Lode Mining Region of California - Unavailable since 1900, this publication includes details on the gold mines of California's famous Mother Lode gold mining area. Included are details on the geology, history and important gold mines of the region, as well as insights into historic mining methods, mine timbering, mining machinery, mining bell signals and other details on how these mines operated. Also included are insights into the gold mines of the California Mother Lode that were in operation during the first sixty years of California's mining history. 8.5" X 11", 176 ppgs. Retail Price: $14.99

Lode Gold of the Klamath Mountains of Northern California and South West Oregon - Unavailable since 1971, this publication was originally compiled by Preston E. Hotz and includes details on the lode mining districts of Oregon and California's Klamath Mountains. Included are details on the geology, history and important lode mines of the French Gulch, Deadwood, Whiskeytown, Shasta, Redding, Muletown, South Fork, Old Diggings, Dog Creek (Delta), Bully Choop (Indian Creek), Harrison Gulch, Hayfork, Minersville, Trinity Center, Canyon Creek, East Fork, New River, Denny, Liberty (Black Bear), Cecilville, Callahan, Yreka, Fort Jones and Happy Camp mining districts in California, as well as the Ashland, Rogue River, Applegate, Illinois River, Takilma, Greenback, Galice, Silver Peak, Myrtle Creek and Mule Creek districts of South Western Oregon. Also included are insights into the mineralization and other characteristics of this important mining region. 8.5" X 11", 100 ppgs. Retail Price: $10.99

Mines and Mineral Resources of Shasta County, Siskiyou County, Trinity County: California - Unavailable since 1915, this publication was originally compiled by the California State Mining Bureau and includes details on the gold mines of this area of Northern California. Also included are insights into the mineralization and other characteristics of this important mining region, as well as the location of historic gold mines. 8.5" X 11", 204 ppgs. Retail Price: $19.99

Geology of the Yreka Quadrangle, Siskiyou County, California - Unavailable since 1977, this publication was originally compiled by Preston E. Hotz and includes details on the geology of the Yreka Quadrangle of Siskiyou County, California. Also included are insights into the mineralization and other characteristics of this important mining region. **8.5" X 11", 78 ppgs. Retail Price: $7.99**

Mines of San Diego and Imperial Counties, California - Originally published in 1914, this important publication on California Mining has not been available for a century. This publication includes important information on the early gold mines of San Diego and Imperial County, which were some of the first gold fields mined in California by early Spanish and Mexican miners before the 49ers came on the scene. Included are not only details on early mining methods in the area, production statistics and geological information, but also the location of the early gold mines that helped make California "The Golden State". Also included are details on the mining of other minerals such as silver, lead, zinc, manganese, tungsten, vanadium, asbestos, barite, borax, cement, clay, dolomite, fluospar, gem stones, graphite, marble, salines, petroleum, stronium, talc and others. **8.5" X 11", 116 ppgs. Retail Price: $12.99**

Mines of Sierra County, California - Unavailable since 1920, this publication was originally compiled by the California State Mining Bureau and includes details on the gold mines of Sierra County, California. Also included are insights into the mineralization and other characteristics of this important mining region, as well as the location of historic gold mines. **8.5" X 11", 156 ppgs. Retail Price: $19.99**

Mines of Plumas County, California - Unavailable since 1918, this publication was originally compiled by the California State Mining Bureau and includes details on the gold mines of Plumas County, California. Also included are insights into the mineralization and other characteristics of this important mining region, as well as the location of historic gold mines. **8.5" X 11", 200 ppgs. Retail Price: $19.99**

Mines of El Dorado, Placer, Sacramento and Yuba Counties, California - Originally published in 1917, this important publication on California Mining has not been available for nearly a century. This publication includes important information on the early gold mines of El Dorado County, Placer County, Sacramento County and Yuba County, which were some of the first gold fields mined by the Forty-Niners during the California Gold Rush. Included are not only details on early mining methods in the area, production statistics and geological information, but also the location of the early gold mines that helped make California "The Golden State". Also included are insights into the early mining of chrome, copper and other minerals in this important mining area. **8.5" X 11", 204 ppgs. Retail Price: $19.99**

Mines of Los Angeles, Orange and Riverside Counties, California - Originally published in 1917, this important publication on California Mining has not been available for nearly a century. This publication includes important information on the early gold mines of Los Angeles County, Orange County and Riverside County, which were some of the first gold fields mined in California by early Spanish and Mexican miners before the 49ers came on the scene. Included are not only details on early mining methods in the area, production statistics and geological information, but also the location of the early gold mines that helped make California "The Golden State". **8.5" X 11", 146 ppgs. Retail Price: $12.99**

Mines of San Bernadino and Tulare Counties, California - Originally published in 1917, this important publication on California Mining has not been available for nearly a century. This publication includes important information on the early gold mines of San Bernadino and Tulare County, which were some of the first gold fields mined in California by early Spanish and Mexican miners before the 49ers came on the scene. Included are not only details on early mining methods in the area, production statistics and geological information, but also the location of the early gold mines that helped make California "The Golden State". Also included are details on the mining of other minerals such as copper, iron, lead, zinc, manganese, tungsten, vanadium, asbestos, barite, borax, cement, clay, dolomite, fluospar, gem stones, graphite, marble, salines, petroleum, stronium, talc and others. **8.5" X 11", 200 ppgs. Retail Price: $19.99**

Chromite Mining in The Klamath Mountains of California and Oregon - Unavailable since 1919, this publication was originally compiled by J.S. Diller of the United States Department of Geological Survey and includes details on the chromite mines of this area of Northern California and Southern Oregon. Also included are insights into the mineralization and other characteristics of this important mining region, as well as the location of historic mines. Also included are insights into chromite mining in Eastern Oregon and Montana. **8.5" X 11", 98 ppgs. Retail Price: $9.99**

Mines and Mining in Amador, Calaveras and Tuolumne Counties, California - Unavailable since 1915, this publication was originally compiled by William Tucker and includes details on the mines and mineral resources of this important California mining area. Included are details on the geology, history and important gold mines of the region, as well as insights into other local mineral resources such as asbestos, clay, copper, talc, limestone and others. Also included are insights into the mineralization and other characteristics of this important portion of California's Mother Lode mining region. **8.5" X 11", 198 ppgs. Retail Price: $14.99**

The Cerro Gordo Mining District of Inyo County California - Unavailable since 1963, this publication was originally compiled by the United States Department of Interior. Included are insights into the mineralization and other characteristics of this important mining region of Southern California. Topics include the mining of gold and silver in this important mining district in Inyo County, California, including details on the history, production and locations of the Cerro Gordo Mine, the Morning Star Mine, Estelle Tunnel, Charles Lease Tunnel, Ignacio, Hart, Crosscut Tunnel, Sunset, Upper Newtown, Newtown, Ella, Perseverance, Newsboy, Belmont and other silver and gold mines in the Cerro Gordo Mining District. This volume also includes important insights into the fossil record, geologic formations, faults and other aspects of economic geology in this California mining district. **8.5″ X 11″, 104 ppgs.** Retail Price: $10.99

Mining in Butte, Lassen, Modoc, Sutter and Tehama Counties of California - Unavailable since 1917, this publication was originally compiled by the United States Department of Interior. Included are insights into the mineralization and other characteristics of this important mining region of California. Topics include the mining of asbestos, chromite, gold, diamonds and manganese in Butte County, the mining of gold and copper in the Hayden Hill and Diamond Mountain mining districts of Lassen County, the mining of coal, salt, copper and gold in the High Grade and Winters mining districts of Modoc County, gold mining in Sutter County and the mining of gold, chromite, manganese and copper in Tehama County. This volume also includes the production records and locations of numerous mines in this important mining region. **8.5″ X 11″, 114 ppgs. Retail Price: $11.99**

Mines of Trinity County California - Originally published in 1965, this important publication on California Mining has not been available for nearly fifty years. This publication includes important information on mines and mining in Trinity County, California, as well insights into the mineralization and geology of this important mining area in Northern California. Included are extensive details on hardrock and placer gold mines and prospects, including charts showing the locations of these historic mines.. **8.5″ X 11″, 144 ppgs. Retail Price: $12.99**

Mines of Kern County California - Originally published in 1962, this important publication on California Mining has not been available for nearly fifty years. This publication includes important information on mines and mining in Kern County, California, as well insights into the mineralization and geology of this important mining area in California. Included are extensive details on hardrock and placer gold mines and prospects, including charts showing the locations of these historic mines. **8.5″ X 11″, 398 ppgs. Retail Price: $24.99**

Mines of Calaveras County California - Originally published in 1962, this important publication on California Mining has not been available for nearly fifty years. This publication includes important information on mines and mining in Calaveras County, California, as well insights into the mineralization and geology of this important mining area in Northern California. Included are extensive details on hardrock and placer gold mines and prospects, including charts showing the locations of these historic mines. **8.5″ X 11″, 236 ppgs. Retail Price: $19.99**

Lode Gold Mining in Grass Valley California - Unavailable since 1940, this publication was originally compiled by the United States Department of Interior. Included are insights into the gold mineralization and other characteristics of this important mining region of Nevada County, California. This volume also includes important insights into the geologic formations, faults and other aspects of economic geology in this California mining district. Of particular interest are the fine details on many hardrock gold mines in the area, including their locations, histories, development and mineralization. Some of the mines featured include the Gold Hill Mine, Massachusetts Hill, Boundary, Peabody, Golden Center, North Star, Omaha, Lone Jack, Homeward Bound, Hartery, Wisconsin, Allison Ranch, Phoenix, Kate Hayes, W.Y.O.D., Empire, Rich Hill, Daisy Hill, Orleans, Sultana, Centennial, Conlin, Ben Franklin, Crown Point and many others. **8.5″ X 11″, 148 ppgs. Retail Price: $12.99**

Lode Mining in the Alleghany District of Sierra County California - Unavailable since 1913, this publication was originally compiled by the United States Department of Interior. Included are insights into the mineralization and other characteristics of this important mining region of Sierra County. Included are details on the history, production and locations of numerous hardrock gold mines in this famous California area, including the Tightner Mine, Minnie D., Osceola, Eldorado, Twenty One, Sherman, Kenton, Oriental, Rainbow, Plumbago, Irelan, Gold Canyon, North Fork, Federal, Kate Hardy and others. This volume also includes important insights into the fossil record, geologic formations, faults and other aspects of economic geology in this California mining district. **8.5″ X 11″, 48 ppgs. Retail Price: $7.99**

Six Months In The Gold Mines During The California Gold Rush - Unavailable since 1850, this important work is a first hand account of one "49'ers" personal experience during the great California Gold Rush, shedding important light on one of the most exciting periods in the history of not only California, but also the world. Compiled from journals written between 1847 and 1849 by E. Gould Buffum, a native of New York, "Six Months In The Gold Mines During The California Gold Rush" offers a rare look into the day to day lives of the people who came to California to work in her gold mines when the state was still a great frontier. **8.5″ X 11″, 290 ppgs. Retail Price: $19.99**

Quartz Mines of the Grass Valley Mining District of California - Unavailable since 1867, this important publication has not been available since those days. This rare publication offers a short dissertation on the early hardrock mines in this important mining district in the California Mother Lode region between the 1850's and 1860's. Also included are hard to find details on the mineralization and locations of these mines, as well as how they were operated in those day. **8.5" X 11", 44 ppgs. Retail Price: $8.99**

Gold Rush on the Feather River - First published in 1924, this short publication by G.C. Mansfield sheds important light on the early history of gold mining on the Feather River. Included are rare insights into the first decade of gold mining and the early mining camps of the Feather River during the 1850's. **64 ppgs., 9.99**

The Bodie Mining District of California - First published in 1986, it has been unavailable since those days and sheds important light on this famous mining area. Included are the history, characteristics and locations of numerous old mines around the ghost town of Bodie.
64 ppgs, 8.99

Geology and Mineral Resources of the Gasquet Quadrangle of California-Oregon - First published in 1953, it has been unavailable for over a century and sheds important light on the geological features and mineral resources of this portion of Northern California and Southern Oregon.
80 ppgs, 9.99

Alaska Mining Books

Ore Deposits of the Willow Creek Mining District, Alaska - Unavailable since 1954, this hard to find publication includes valuable insights into the Willow Creek Mining District near Hatcher Pass in Alaska. The publication includes insights into the history, geology and locations of the well known mines in the area, including the Gold Cord, Independence, Fern, Mabel, Lonesome, Snowbird, Schroff-O'Neil, High Grade, Marion Twin, Thorpe, Webfoot, Kelly-Willow, Lane, Holland and others. **8.5" X 11", 96 ppgs. Retail Price: $9.99**

The Juneau Gold Belt of Alaska - Unavailable since 1906, this hard to find publication includes valuable insights into the gold mines around Juneau, Alaska. The publication includes important details into the history, geology and locations of the well known gold mines and prospects in the area, including those around Windham Bay, Holkham Bay, Port Snettisham, on Grindstone and Rhine Creeks, Gold Creek, Douglas Island, Salmon Creek, Lemon Creek, Nugget Creek, from the Mendenhall River to Berners Bay, McGinnis Creek, Montana Creek, Peterson Creek, Windfall Creek, the Eagle River, Yankee Basin, Yankee Curve, Kowee Creek and elsewhere. Not only are gold placer mines included, but also hardrock gold mines. **8.5" X 11", 224 ppgs. Retail Price: $19.99**

Mining in the Jumbo Basin of Alaska - Unavailable since 1953, this hard to find publication includes valuable insights into the mines and geology of the Jumbo Basin. The publication includes important details into the history, geology and locations of the well known gold mines and prospects in the famous Jumbo Basin Mining Region of Alaska.
72 ppgs, 9.99

The Rampart Placer Gold Region of Alaska - Unavailable since 1906, this hard to find publication includes valuable insights into the placer gold mines of the Rampart Mining Region. The publication includes important details into the history, geology and locations of the well known gold mines and prospects in the famous Rampart Mining Region of Alaska.
78 ppgs, 10.99

Arizona Mining Books

Mines and Mining in Northern Yuma County Arizona - Originally published in 1911, this important publication on Arizona Mining has not been available for over a hundred years. Included are rare insights into the gold, silver, copper and quicksilver mines of Yuma County, Arizona together with hard to find maps and photographs. Some of the mines and mining districts featured include the Planet Copper Mine, Mineral Hill, the Clara Consolidated Mine, Viati Mine, Copper Basin prospect, Bowman Mine, Quartz King, Billy Mack, Carnation, the Wardwell and Osbourne, Valensuella Copper, the Mariquita, Colonial Mine, the French American, the New York-Plomosa, Guadalupe, Lead Camp, Mudersbach Copper Camp, Yellow Bird, the Arizona Northern (Salome Strike), Bonanza (Harqua Hala), Golden Eagle, Hercules, Socorro and others. **8.5" X 11", 144 ppgs. Retail Price: $11.99**

The Aravaipa and Stanley Mining Districts of Graham County Arizona - Originally published in 1925, this important publication on Arizona Mining has not been available for nearly ninety years. Included are rare insights into the gold and silver mines of these two important mining districts, together with hard to find maps. **8.5" X 11", 140 ppgs. Retail Price: $11.99**

Gold in the Gold Basin and Lost Basin Mining Districts of Mohave County, Arizona - This volume contains rare insights into the geology and gold mineralization of the Gold Basin and Lost Basin Mining Districts of Mohave County, Arizona that will be of benefit to miners and prospectors. Also included is a significant body of information on the gold mines and prospects of this portion of Arizona. This volume is lavishly illustrated with rare photos and mining maps. **8.5" X 11", 188 ppgs. Retail Price: $19.99**

Mines of the Jerome and Bradshaw Mountains of Arizona - This important publication on Arizona Mining has not been available for ninety years. This volume contains rare insights into the geology and ore deposits of the Jerome and Bradshaw Mountains of Arizona that will be of benefit to miners and prospectors who work those areas. Included is a significant body of information on the mines and prospects of the Verde, Black Hills, Cherry Creek, Prescott, Walker, Groom Creek, Hassayampa, Bigbug, Turkey Creek, Agua Fria, Black Canyon, Peck, Tiger, Pine Grove, Bradshaw, Tintop, Humbug and Castle Creek Mining Districts. This volume is lavishly illustrated with rare photos and mining maps. **8.5" X 11", 218 ppgs. Retail Price: $19.99**

The Ajo Mining District of Pima County Arizona - This important publication on Arizona Mining has not been available for nearly seventy years. This volume contains rare insights into the geology and mineralization of the Ajo Mining District in Pima County, Arizona and in particular the famous New Cornelia Mine. **8.5" X 11", 126 ppgs. Retail Price: $11.99**

Mining in the Santa Rita and Patagonia Mountains of Arizona - Originally published in 1915, this important publication on Arizona Mining has not been available for nearly a century. Included are rare insights into hundreds of gold, silver, copper and other mines in this famous Arizona mining area. Details include the locations, geology, history, production and other facts of the mines of this region. **8.5" X 11", 394 ppgs. Retail Price: $24.99**

Mining in the Bisbee Quadrangle of Arizona - Originally published in 1906, this important publication on Arizona Mining has not been available for nearly a century. Included are rare insights into hundreds of gold, silver, copper and other mines in this famous Arizona mining area. Details include the locations, geology, history, production and other facts of the mines of this important mining region. **8.5" X 11", 188 ppgs. Retail Price: $14.99**

Placer Gold Mining in Arizona - Unavailable since 1922, this hard to find publication includes valuable insights into the placer gold mines of the Arizona. Originally released as "Placer Gold of Arizona", despite its small size, this publication includes important details into the history, geology and locations of the well known placer gold mines and prospects in the State of Arizona. **48 ppgs, 8.99**

Gold and Copper Mining near Payson, Arizona - Written in 1915, this hard to find publication includes valuable insights into the gold and copper mining industry of Arizona. Highlighted here are the gold and copper mines near Payson, Arizona. **68 ppgs, 8.99**

Lode Gold Mining in Arizona - Unavailable since 1934, this hard to find publication, originally released as "Arizona Lode Gold Mines and Gold Mining" includes valuable insights into the gold mining industry of Arizona. Included are valuable insights into over 150 hardrock gold mines in over 30 different mining districts in Arizona. **278 ppgs, 21.99**

Mining in the Dragoon Quadrangle of Cochise County, Arizona - Unavailable since 1964, this hard to find publication includes valuable insights into the mines of the Dragoon Quadrangle Mining Region. The publication includes important details into the history, geology and locations of the well known mines and prospects in this famous mining region of Arizona. **224 ppgs., 19.99**

Directory of Operating Mines in Arizona in 1915 - Unavailable since 1916, this hard to find publication includes valuable insights into the mines of Arizona. This small publication includes a complete list of the mines that were operating in the State of Arizona during 1915 and includes details such as general location, owners and some basic facts about each mining operation. **52 ppgs. 8.99**

Arizona Ore Deposits - Unavailable since 1938, this hard to find publication includes valuable insights into some ore deposits of Arizona. Included are valuable insights into the formation and characteristics of valuable ore deposits in the Jerome, Miami, Inspiration, Clifton, Morenci, Ray, Ajo, Eureka, Tombstone and Magma mining districts. Included are details into some of the major gold, silver and copper mines of these important Arizona mining areas. **160 ppgs, 14.99**

Montana Mining Books

A History of Butte Montana: The World's Greatest Mining Camp - First published in 1900 by H.C. Freeman, this important publication sheds a bright light on one of the most important mining areas in the history of The West. Together with his insights, as well as rare photographs of the periods, Harry Freeman describes Butte and its vicinity from its early beginnings, right up to its flush years when copper flowed from its mines like a river. At the time of publication, Butte, Montana was known worldwide as "The Richest Mining Spot On Earth" and produced not only vast amounts of copper, but also silver, gold and other metals from its mines. Freeman illustrates, with great detail, the most important mines in the vicinity of Butte, providing rare details on their owners, their history and most importantly, how the mines operated and how their treasures were extracted. Of particular interest are the dozens of rare photographs that depict mines such as the famous Anaconda, the Silver Bow, the Smoke House, Moose, Paulin, Buffalo, Little Minah, the Mountain Consolidated, West Greyrock, Cora, the Green Mountain, Diamond, Bell, Parnell, the Neversweat, Nipper, Original and many others. **8.5" X 11", 142 ppgs. Retail Price: $12.99**

The Butte Mining District of Montana - This important publication on Montana Mining has not been available for over a century. Included are rare insights into the gold, copper and silver mines of Butte, Montana together with hard to find maps and photographs. Some of the topics include the early history of gold, silver and copper mining in the Butte area, insight into the geology of its mining areas, the local distribution of gold, silver and copper ores, as well their composition and how to identify them. Also included are detailed facts about the mines in the Butte Mining District, including the famous Anaconda Mine, Gagnon, Parrot, Blue Vein, Moscow, Poulin, Stella, Buffalo, Green Mountain, Wake Up Jim, the Diamond-Bell Group, Mountain Consolidated, East Greyrock, West Greyrock, Snowball, Corra, Speculator, Adirondack, Miners Union, the Jessie-Edith May Group, Otisco, Iduna, Colorado, Lizzie, Cambers, Anderson, Hesperus, Preferencia and dozens of others. **8.5" X 11", 298 ppgs. Retail Price: $24.99**

Mines of the Helena Mining Region of Montana - This important publication on Montana Mining has not been available for over a century. Included are rare insights into the gold, copper and silver mines of the vicinity of Helena, Montana, including the Marysville Mining District, Elliston Mining District, Rimini Mining District, Helena Mining District, Clancy Mining District, Wickes Mining District, Boulder and Basin Mining Districts and the Elkhorn Mining District. Some of the topics include the early history of gold, silver and copper mining in the Helena area, insight into the geology of its mining areas, the local distribution of gold, silver and copper ores, as well their composition and how to identify them. Also included are detailed facts, history, geology and locations of over one hundred gold, silver and copper mines in the area . **8.5" X 11", 162 ppgs, Retail Price: $14.99**

Mines and Geology of the Garnet Range of Montana - This important publication on Montana Mining has not been available for over a century. Included are rare insights into the gold, copper and silver mines of the vicinity of this important mining area of Montana. Some of the topics include the early history of gold, silver and copper mining in the Garnet Mountains, insight into the geology of its mining areas, the local distribution of gold, silver and copper ores, as well their composition and how to identify them. Also included are detailed facts, history, geology and locations of numerous gold, silver and copper mines in the area . **8.5" X 11", 100 ppgs, Retail Price: $11.99**

Mines and Geology of the Philipsburg Quadrangle of Montana - This important publication on Montana Mining has not been available for over a century. Included are rare insights into the gold, copper and silver mines of the vicinity of this important mining area of Montana. Some of the topics include the early history of gold, silver and copper mining in the Philipsburg Quadrangle, insight into the geology of its mining areas, the local distribution of gold, silver and copper ores, as well their composition and how to identify them. Also included are detailed facts, history, geology and locations of over one hundred gold, silver and copper mines in the area **8.5" X 11", 290 ppgs, Retail Price: $24.99**

Geology of the Marysville Mining District of Montana - Included are rare insights into the mining geology of the Marysville Mining District. Some of the topics include the early history of gold, silver and copper mining in the area, insight into the geology of its mining areas, the local distribution of gold, silver and copper ores, as well their composition and how to identify them. Also included are detailed facts, history, geology and locations of gold, silver and copper mines in the area **8.5" X 11", 198 ppgs, Retail Price: $19.99**

The Geology and Mines of Northern Idaho and North Western Montana- See listing under Idaho.

The History of Gold Dredging in Montana - Unavailable since 1916, this important publication was originally published by the Us Bureau of Mines and has been unavailable for a century. A century and more ago, giant dredging machines dug in Montana's rivers and creeks in search of illusive golden riches. First appearing in California in the 1850's, gold dredges finally reached their peak of development in Siberia and New Zealand before becoming popular again in the United States. This book offers a unique historical perspective on the gold dredges that once operated in Montana. This book on Montana mining history is lavishly illustrated with dozens of rare historic photos gold dredges that once operated in Montana, as well as hard to locate plans on how these dredges were designed. 120 ppgs., 11.99

Nevada Mining Books

The Bull Frog Mining District of Nevada - Unavailable since 1910, this publication was originally compiled by the United States Department of Interior. This volume also includes important insights into the geologic formations, faults and other aspects of economic geology in this Nevada mining district. Of particular interest are the fine details on many mines in the area, including their locations, histories, development and mineralization. Some of the mines featured include the National Bank Mine, Providence, Gibraltor, Tramps, Denver, Original Bullfrog, Gold Bar, Mayflower, Homestake-King and other mines and prospects. **8.5" X 11", 152 ppgs, Retail Price: $14.99**

History of the Comstock Lode - Unavailable since 1876, this publication was originally released by John Wiley & Sons. This volume also includes important insights into the famous Comstock Lode of Nevada that represented the first major silver discovery in the United States. During its spectacular run, the Comstock produced over 192 million ounces of silver and 8.2 million ounces of gold. Not only did the Comstock result in one of the largest mining rushes in history and yield immense fortunes for its owners, but it made important contributions to the development of the State of Nevada, as well as neighboring California. Included here are important details on not only the early development and history of the Comstock, but also rare early insight into its mines, ore and its geology.**8.5" X 11", 244 ppgs, Retail Price: $19.99**

The Pioche Mining District of Nevada - First published in 1932, it has been unavailable for over a century and sheds important light on the mining history of Nevada. Some of the topics include the history of mining in this district, as well as the characteristics of its mineral and ore deposits. Also included are insights into the history, production, characteristics and locations of numerous mines in the area. Some of the mines include the Combined Metals, Pioche, Ely Valley, No. 10, Poorman, Wide Awake, Alps, Prince, Virginia Louise, Half Moon, Abe Lincoln, Fairview, Bristol Silver, National, Vesuvius, Inman, Tempest, Hillside, Jackrabbit, Lucky Star, Fortuna, Mendha, Manhattan, Hamburg, Comet, Lyndon and others. 108 ppgs 10.99

The Yerington Mining District of Nevada - First published in 1932, it has been unavailable for over a century and sheds important light on the mining history of Nevada. Some of the topics include the history of mining in this district, as well as the characteristics of its mineral and ore deposits. Also included are insights into the history, production, characteristics and locations of numerous mines in the area. Some of the mines include the Bluestone, Mason Valley, Malachite, McConnell, Greenwood, Western Nevada, Ludwig, Douglas Hill, Casting Copper, Montana-Yerington, Empire, Jim Beatty, Terry and McFarland, Blue Jay and others. 92 ppgs, 10.99

The Genesis of the Ores of Tonopah Nevada - Unavailable since 1918, this hard to find publication includes valuable insights into the gold mines around Tonopah, Nevada. The publication includes important details into the geology of mines in the Tonopah Mining District of Nevada. 90 ppgs, 10.99

Mining Camps of Elko, Lander and Eureka Counties Nevada - Unavailable since 1910, this hard to find publication includes valuable insights into the mining camps of Elko, Lander and Eureka Counties, Nevada. The publication includes important details into the history of mines and mining in these three Nevada counties. 154 ppgs, 12.99

Ore Deposits of the Bullfrog Quadrangle - Unavailable since 1964 and released as "Geology of Bullfrog Quadrangle and Ore Deposits Related to Bullfrog Hills Caldera, Nye County, Nevada and Inyo County, California". The publication includes important details into the geology of mines in the Bullfrog Quadrangle of Nye County, Nevada and Inyo County, California. 52 ppgs, 9.99

Mining in Eureka County Nevada - Unavailable since 1879, this hard to find publication includes valuable insights into the early mining history off Eureka County, Nevada. The publication includes important details into the early history of the mines of Eureka County, as well as their development, production and how their ores were treated. Also included are details on the 1872 Mining Act, as well as the local rules, regulations and customs of the miners in Eureka County.134 ppgs, 12.99

Colorado Mining Books

Ores of The Leadville Mining District - Unavailable since 1926, this publication was originally compiled by the United States Department of Interior. This volume also includes important insights into the ores and mineralization of the Leadville Mining District in Colorado. Topics include historic ore prospecting methods, local geology, insights into ore veins and stockworks, the local trend and distribution of ore channels, reverse faults, shattered rock above replacement ore bodies, mineral enrichment in oxidized and sulphide zones and more. **8.5" X 11", 66 ppgs, Retail Price: $8.99**

Mining in Colorado - Unavailable since 1926, this publication was originally compiled by the United States Department of Interior. This volume also includes important insights into the mining history of Colorado from its early beginnings in the 1850's right up to the mid 1920's. Not only is Colorado's gold mining heritage included, but also its silver, copper, lead and zinc mining industry. Each mining area is treated separately, detailing the development of Colorado's mines on a county by county basis. **8.5" X 11", 284 ppgs, Retail Price: $19.99**

Gold Mining in Gilpin County Colorado - Unavailable since 1876, this publication was originally compiled by the Register Steam Printing House of Central City, Colorado. A rare glimpse at the gold mining history and early mines of Gilpin County, Colorado from their first discovery in the 1850's up to the "flush years" of the mid 1870's. Of particular interest is the history of the discovery of gold in Gilpin County and details about the men who made those first strikes. Special focus is given to the early gold mines and first mining districts of the area, many of which are not detailed in other books on Colorado's gold mining history. **8.5" X 11", 156 ppgs, Retail Price: $12.99**

Mining in the Gold Brick Mining District of Colorado - Important insights into the history of the Gold Brick Mining District, as well as its local geography and economic geology. Also included are the histories and locations of historic mines in this important Colorado Mining District, including the Cortland, Carter, Raymond, Gold Links, Sacramento, Bassick, Sandy Hook, Chronicle, Grand Prize, Chloride, Granite Mountain, Lucille, Gray Mountain, Hilltop, Maggie Mitchell, Silver Islet, Revenue, Roosevelt, Carbonate King and others. In addition to hardrock mining, are also included are details on gold placer mining in this portion of Colorado. **8.5" X 11", 140 ppgs, Retail Price: $12.99**

Ore Deposits of the London Fault of Colorado - First published in 1941, it has been unavailable since those days and sheds important light on the mines and mineral deposits of the London Fault in Central Colorado's Alma Mining District. This publication sheds important light on the gold veins and lead-silver deposits of the Alma Mining District. Included are geologic details on the London Mine, American Mine, Havigorst Tunnel, Ophir Mine, Mosher Tunnel, London-Butte Mine, Venture Shaft, Hard-To-Beat Mine, Oliver Twist Tunnel, Sacramento Mine, Mudsill Mine, Sherwood Mine, Wagner, Barcoe Tunnel and other mines in this important mining region. 110 ppgs., 10.99

The Mines of Colorado - First published in 1867, it has been unavailable since those days and sheds important light on Colorado's early mining history. Written shortly after the events took place, this publication sheds important light on the Pike's Peak Gold Rush, the discovery of gold on Ralston Creek and Dry Creek in the 1850's, as well as details on the first wave of miners into Colorado and their trials and tribulations as they crossed the Great Plains. Also included are details on early discoveries of lode gold in the mountainous regions of Colorado, details on the early mines hardrock and placer mines, and much more. It is a veritable treasure trove on Colorado's early mining history and will be of great importance to anyone who is interested in the mining of gold or other minerals in Colorado, as well as those interested in the history of the state. 478 ppgs., 29.99

The La Plata Mining District of Colorado - Originally titled "Geology and Ore Deposits in the Vicinity of the La Plata District of Colorado" and first published in 1949, it has been unavailable since those days and sheds important light on the mines and mineral deposits of the La Plata Mining District of Colorado. 214 ppgs., 19.99

Washington Mining Books

The Republic Mining District of Washington - Unavailable since 1910, this important publication was originally published by the Washington Geologic Survey and has been unavailable for a century. Topics include the geology, rock formations and the formation of ore deposits in this important mining area of Washington State. Also included are hard to find details on the geology, history and locations of dozens of mines in the area. Some of the mines featured include the New Republic Mine, Ben Hur, Morning Glory, the South Republic Mine, Quilp, Surprise, Black Tail, Lone Pine, San Poil, Mountain Lion, Tom Thumb, Elcaliph and many others. **8.5" X 11", 94 ppgs, Retail Price: $10.99**

The Myers Creek and Nighthawk Mining Districts of Washington - Unavailable since 1911, this important publication was originally published by the Washington Geologic Survey and has been unavailable for a century. Topics include the geology, rock formations and the formation of ore deposits in these important mining areas of Washington State. Also included are hard to find details on the geology, history and locations of dozens of mines in the area. Some of the mines featured include the Grant Mine, Monterey, Nip and Tuck, Myers Creek, Number Nine, Neutral, Rainbow, Aztec, Crystal Butte, Apex, Butcher Boy, Molson, Mad River, Olentangy, Delate, Kelsey, Golden Chariot, Okanogan, Ohio, Forty-Ninth Parallel, Nighthawk, Favorite, Little Chopaka, Summit, Number One, California, Peerless, Caaba, Prize Group, Ruby, Mountain Sheep, Golden Zone, Rich Bar, Similkameen, Kimberly, Triune, Hiawatha, Trinity, Hornsilver, Maquae, Bellevue, Bullfrog, Palmer Lake, Ivanhoe, Copper World and many others. **8.5" X 11", 136 ppgs, Retail Price: $12.99**

The Blewett Mining District of Washington - Unavailable since 1911, this important publication was originally published by the Washington Geologic Survey and has been unavailable for a century. Topics include the geology, rock formations and the formation of ore deposits in this important mining area of Washington State. Also included are hard to find details on the geology, history and locations of dozens of mines in the area. Some of the mines featured include the Washington Meteor, Alta Vista, Pole Pick, Blinn, North Star, Golden Eagle, Tip Top, Wilder, Golden Guinea, Lucky Queen, Blue Bell, Prospect, Homestake, Lone Rock, Johnson, and others. **8.5" X 11", 134 ppgs, Retail Price: $12.99**

Silver Mining In Washington - Unavailable since 1955, this important publication was originally published by the Washington Geologic Survey. Featured are the hard to find locations and details pertaining to Washington's silver mines. **8.5" X 11", 180 ppgs, Retail Price: $15.99**

The Mines of Snohomish County Washington - Unavailable since 1942, this important publication was originally published by the Washington Geologic Survey and has been unavailable for seventy years. Featured are details on a large number of gold, silver, copper, lead and other metallic mineral mines. Included are the locations of each historic mine, along with information on the commodity produced. **8.5" X 11", 98 ppgs, Retail Price: $10.99**

The Mines of Chelan County Washington - Unavailable since 1943, this important publication was originally published by the Washington Geologic Survey and has been unavailable for seventy years. Featured are details on a large number of gold, silver, copper, lead and other metallic mineral mines. Included are the locations of each historic mine, along with information on the commodity. **8.5" X 11", 88 ppgs, Retail Price: $9.99**

Metal Mines of Washington - Unavailable since 1921, this important publication was originally published by the Washington Geologic Survey and has been unavailable for nearly ninety years. Widely considered a masterpiece on the Washington Mining Industry, "Metal Mines of Washington" sheds light on the important details of Washington's early mining years. Featured are details on hundreds of gold, silver, copper, lead and other metallic mineral mines. Included are hard to find details on the mineral resources of this state, as well as the locations of historic mines. Lavishly illustrated with maps and historic photos and complete with a glossary to explain any technical terms found in the text, this is one of the most important works on mining in the State of Washington. No prospector or miner should be without it if they are interested in mining in Washington. **8.5" X 11", 396 ppgs, Retail Price: $24.99**

Gem Stones In Washington - Unavailable since 1949, this important publication was originally published by the Washington Geologic Survey and has been unavailable since first published. Included are details on where to find naturally occurring gem stones in the State of Washington, including quartz crystal, amethyst, smoky quartz, milky quartz, agates, bloodstone, carnelian, chert, flint, jasper, onyx, petrified wood, opal, fire opal, hyalite and others. **8.5" X 11", 54 ppgs, Retail Price: $8.99**

The Covada Mining District of Washington - Unavailable since 1913, this important publication was originally published by the Washington Geologic Survey and has been unavailable for a century. Topics include the geology, rock formations and the formation of ore deposits in this important mining area of Washington State. Also included are hard to find details on the geology, history and locations of dozens of mines in the area. Some of the mines featured include the Admiral, Advance, Algonkian, Big Bug, Big Chief, Big Joker, Black Hawk, Black Tail, Black Thorn, Captain, Cherokee Strip, Colorado, Dan Patch, Dead Shot, Etta, Good Ore, Greasy Run, Great Scott, Idora, IXL, Jay Bird, Kentucky Bell, King Solomon, Laurel, Laura S, Little Jay, Meteor, Neglected, Northern Light, Old Nell, Plymouth Rock, Polaris, Quandary, Reserve, Shoo Fly, Silver Plume, Three Pines, Vernie, White Rose and dozens of others. **8.5" X 11", 114 ppgs, Retail Price: $10.99**

The Index Mining District of Washington - Unavailable since 1912, this important publication was originally published by the Washington Geologic Survey and has been unavailable for a century. Topics include the geology, rock formations and the formation of ore deposits in this important mining area of Washington State. Also included are hard to find details on the geology, history and locations of dozens of mines in the area. Some of the mines featured include the Sunset, Non-Pareil, Ethel Consolidated, Kittaning, Merchant, Homestead, Co-operative, Lost Creek, Uncle Sam, Calumet, Florence-Rae, Bitter Creek, Index Peacock, Gunn Peak, Helena, North Star, Buckeye. Copper Bell, Red Cross and others. 8.5" X 11", 114 ppgs, **Retail Price: $11.99**

Mining & Mineral Resources of Stevens County Washington - Unavailable since 1920, this important publication was originally published by the Washington Geologic Survey and has been unavailable for a century. Topics include the geology, rock formations and the formation of ore deposits in these important mining areas of Washington State. Also included are hard to find details on the geology, history and locations of hundreds of mines in the area. 8.5" X 11", 372 ppgs, **Retail Price: $24.99**

The Mines and Geology of the Loomis Quadrangle Okanogan County, Washington - Unavailable since 1972, this important publication was originally published by the Washington Geologic Survey and has been unavailable for a century. Topics include the geology, rock formations and the formation of ore deposits in this important mining area of Washington State. Also included are hard to find details on the geology, history and locations of dozens of gold, copper, silver and other mines in the area. 8.5" X 11", 150 ppgs, **Retail Price: $12.99**

The Conconully Mining District of Okanogan County Washington - Unavailable since 1973, this important publication was originally published by the Washington Geologic Survey and has been unavailable for a century. Topics include the geology, rock formations and the formation of ore deposits in this important mining area of Washington State, which also includes Salmon Creek, Blue Lake and Galena. Also included are hard to find details on the geology, mining history and locations of dozens of mines in the area. Some of the mines include Arlington, Fourth of July, Sonny Boy, First Thought, Last Chance, War Eagle-Peacock, Wheeler, Mohawk, Lone Star, Woo Loo Moo Loo, Keystone, Hughes, Plant-Callahan, Johnny Boy, Leuena, Gubser, John Arthur, Tough Nut, Homestake, Key and many others 8.5" X 11", 68 ppgs, **Retail Price: $8.99**

Wyoming Mining Books

Mining in the Laramie Basin of Wyoming - Unavailable since 1909, this publication was originally compiled by the United States Department of Interior. Also included are insights into the mineralization and other characteristics of this important mining region, especially in regards to coal, limestone, gypsum, bentonite clay, cement, sand, clay and copper. 8.5" X 11", 104 ppgs, **Retail Price: $11.99**

New Mexico Mining Books

The Mogollon Mining District of New Mexico - Unavailable since 1927, this important publication was originally published by the US Department of Interior and has been unavailable for 80 years. Topics include the geology, rock formations and the formation of ore deposits in this important mining area in New Mexico. Of particular focus is information on the history and production of the ore deposits in this area, their form and structure, vein filling, their paragenesis, origins and ore shoots, as well as oxidation and supergene enrichment. Also included are hard to find details, including the descriptions and locations of numerous gold, silver and other types of mines, including the Eureka, Pacific, South Alpine, Great Western, Enterprise, Buffalo, Mountain View, Floride, Gold Dust, Last Chance, Deadwood, Confidence, Maud S., Deep Down, Little Fanney, Trilby, Johnson, Alberta, Comet, Golden Eagle, Cooney, Queen, the Iron Crown, Eberle, Clifton, Andrew Jackson mine, Mascot and others. 8.5" X 11", 144 ppgs, **Retail Price: $12.99**

The Percha Mining District of Kingston New Mexico - Unavailable since 1883, this important publication was originally published by the Kingston Tribune and has been unavailable for over one hundred and thirty five years. Having been written during the earliest years of gold and silver mining in the Percha Mining District, unlike other books on the subject, this work offers the unique perspective of having actually been written while the early mining history of this area was still being made. In fact, the work was written so early in the development of this area that many of the notable mines in the Percha District were less than a few years old and were still being operated by their original discoverers with the same enthusiasm as when they were first located. Included are hard to find details on the very earliest gold and silver mines of this important mining district near Kingston in Sierra County, New Mexico. 8.5" X 11", 68 ppgs, **Retail Price: $9.99**

East Coast Mining Books

<u>The Gold Fields of the Southern Appalachians</u> - Unavailable since 1895, this important publication was originally published by the US Department of Interior and has been unavailable for nearly 120 years. Topics include the geology, rock formations and the formation of ore deposits in this important mining area of the American South. Of particular focus is information on the history and statistics of the ore deposits in this area, their form and structure and veins. Also included are details on the placer gold deposits of the region. The gold fields of the Georgian Belt, Carolinian Belt and the South Mountain Mining District of North Carolina are all treated in descriptive detail. Included are hard to find details, including the descriptions and locations of numerous gold mines in Georgia, North Carolina and elsewhere in the American South. Also included are details on the gold belts of the British Maritime Provinces and the Green Mountains. **8.5" X 11", 104 ppgs, Retail Price: $9.99**

Gold Rush Tales Series

<u>**Millions in Siskiyou County Gold**</u> - In this first volume of the "Gold Rush Tales" series, leading mining historian and editor Kerby Jackson, introduces us to the story of how millions of dollars worth of gold was discovered in Siskiyou County during the California Gold Rush. Lavishly illustrated with photos from the 19th Century, this hard to find information was first published in 1897 and sheds important light onto the gold rush era in Siskiyou County, California and the experiences of the men who dug for the gold and actually found it. **8.5" X 11", 82 ppgs, Retail Price: $9.99**

<u>**The California Rand in the Days of '49**</u> - In this second volume of the "Gold Rush Tales" series, leading mining historian and editor Kerby Jackson, introduces us to four tales from the California Gold Rush. Lavishly illustrated with photos from the 19th Century, this hard to find information was first published in 1890's and includes the stories of "California's Rand", details about Chinese miners, how one early miner named Baker struck it rich and also the story of Alphonzo Bowers, who invented the first hydraulic gold dredge. **8.5" X 11", 54 ppgs, Retail Price: $9.99**

More Mining Books

<u>**Prospecting and Developing A Small Mine**</u> - Topics covered include the classification of varying ores, how to take a proper ore sample, the proper reduction of ore samples, alluvial sampling, how to understand geology as it is applied to prospecting and mining, prospecting procedures, methods of ore treatment, the application of drilling and blasting in a small mine and other topics that the small scale miner will find of benefit. **8.5" X 11", 112 ppgs, Retail Price: $11.99**

<u>**Timbering For Small Underground Mines**</u> - Topics covered include the selection of caps and posts, the treatment of mine timbers, how to install mine timbers, repairing damaged timbers, use of drift supports, headboards, squeeze sets, ore chute construction, mine cribbing, square set timbering methods, the use of steel and concrete sets and other topics that the small underground miner will find of benefit. This volume also includes twenty eight illustrations depicting the proper construction of mine timbering and support systems that greatly enhance the practical usability of the information contained in this small book. **8.5" X 11", 88 ppgs. Retail Price: $10.99**

<u>**Timbering and Mining**</u> - A classic mining publication on Hard Rock Mining by W.H. Storms. Unavailable since 1909, this rare publication provides an in depth look at American methods of underground mine timbering and mining methods. Topics include the selection and preservation of mine timbers, drifting and drift sets, driving in running ground, structural steel in mine workings, timbering drifts in gravel mines, timbering methods for driving shafts, positioning drill holes in shafts, timbering stations at shafts, drainage, mining large ore bodies by means of open cuts or by the "Glory Hole" system, stoping out ore in flat or low lying veins, use of the "Caving System", stoping in swelling ground, how to stope out large ore bodies, Square Set timbering on the Comstock and its modifications by California miners, the construction of ore chutes, stoping ore bodies by use of the "Block System", how to work dangerous ground, information on the "Delprat System" of stoping without mine timbers, construction and use of headframes and much more. This volume provides a reference into not only practical methods of mining and timbering that may be employed in narrow vein mining by small miners today, but also rare insights into how mines were being worked at the turn of the 19th Century. **8.5" X 11", 288 ppgs. Retail Price: $24.99**

A Study of Ore Deposits For The Practical Miner - Mining historian Kerby Jackson introduces us to a classic mining publication on ore deposits by J.P. Wallace. First published in 1908, it has been unavailable for over a century. Included are important insights into the properties of minerals and their identification, on the occurrence and origin of gold, on gold alloys, insights into gold bearing sulfides such as pyrites and arsenopyrites, on gold bearing vanadium, gold and silver tellurides, lead and mercury tellurides, on silver ores, platinum and iridium, mercury ores, copper ores, lead ores, zinc ores, iron ores, chromium ores, manganese ores, nickel ores, tin ores, tungsten ores and others. Also included are facts regarding rock forming minerals, their composition and occurrences, on igneous, sedimentary, metamorphic and intrusive rocks, as well as how they are geologically disturbed by dikes, flows and faults, as well as the effects of these geologic actions and why they are important to the miner. Written specifically with the common miner and prospector in mind, the book will help to unlock the earth's hidden wealth for you and is written in a simple and concise language that anyone can understand. **8.5" X 11", 366 ppgs. Retail Price: $24.99**

Mine Drainage - Unavailable since 1896, this rare publication provides an in depth look at American methods of underground mine drainage and mining pump systems. This volume provides a reference into not only practical methods of mining drainage that may be employed in narrow vein mining by small miners today, but also rare insights into how mines were being worked at the turn of the 19th Century. **8.5" X 11", 218 ppgs. Retail Price: $24.99**

Fire Assaying Gold, Silver and Lead Ores - Unavailable since 1907, this important publication was originally published by the Mining and Scientific Press and was designed to introduce miners and prospectors of gold, silver and lead to the art of fire assaying. Topics include the fire assaying of ores and products containing gold, silver and lead; the sampling and preparation of ore for an assay; care of the assay office, assay furnaces; crucibles and scorifiers; assay balances; metallic ores; scorification assays; cupelling; parting' crucible assays, the roasting of ores and more. This classic provides a time honored method of assaying put forward in a clear, concise and easy to understand language that will make it a benefit to even beginners. **8.5" X 11", 96 ppgs. Retail Price: $11.99**

Methods of Mine Timbering - Originally published in 1896, this important publication on mining engineering has not been available for nearly a century. Included are rare insights into historical methods of timbering structural support that were used in underground metal mines during the California that still have a practical application for the small scale hardrock miner of today. **8.5" X 11", 94 ppgs. Retail Price: $10.99**

The Enrichment of Copper Sulfide Ores - First published in 1913, it has been unavailable for over a century. Topics include the definition and types of ore enrichment, the oxidation of copper ores, the precipitation of metallic sulfides. Also included are the results of dozens of lab experiments pertaining to the enrichment of sulfide ores that will be of interest to the practical hard rock mine operator in his efforts to release the metallic bounty from his mine's ore. **8.5" X 11", 92 ppgs. Retail Price: $9.99**

A Study of Magmatic Sulfide Ores - Unavailable since 1914, this rare publication provides an in depth look at magmatic sulfide ores. Some of the topics included are the definition and classification of magmatic ores, descriptions of some magmatic sulfide ore deposits known at the time of publication including copper and nickel bearing pyrrohitic ore bodies, chalcopyrite-bornite deposits, pyritic deposits, magnetite-ileminite deposits, chromite deposits and magmatic iron ore deposits. Also included are details on how to recognize these types of ore deposits while prospecting for valuable hardrock minerals. **8.5" X 11", 138 ppgs. Retail Price: $11.99**

The Cyanide Process of Gold Recovery - Unavailable since 1894 and released under the name "The Cyanide Process: Its Practical Application and Economical Results", this rare publication provides an in depth look at the early use of cyanide leaching for gold recovery from hardrock mine ores. This volume provides a reference into the early development and use of cyanide leaching to recover gold. **8.5" X 11", 162 ppgs. Retail Price: $14.99**

California Gold Milling Practices - Unavailable since 1895 and released under the name "California Gold Practices", this rare publication provides an in depth look at early methods of milling used to reduce gold ores in California during the late 19th century. This volume provides a reference into the early development and use of milling equipment during the earliest years of the California Gold Rush up to the age of the Industrial Revolution. Much of the information still applies today and will be of use to small scale miners engaging in hardrock mining. **8.5" X 11", 104 ppgs. Retail Price: $10.99**

Leaching Gold and Silver Ores With The Plattner and Kiss Processes - Mining historian Kerby Jackson introduces us to a classic mining publication on the evaluation and examination of mines and prospects by C.H. Aaron. First published in 1881, it has been unavailable for over a century and sheds important light on the leaching of gold and silver ores with the Plattner and Kiss processes. **8.5" X 11", 204 ppgs. Retail Price: $15.99**

The Metallurgy of Lead and the Desilverization of Base Bullion - First published in 1896, it has been unavailable for over a century and sheds important light on the the recovery of silver from lead based ores. Some of the topics include the properties of lead and some of its compounds, lead ores such as galenite, anglesite, cerussite and others, the distribution of lead ores throughout the United States and the sampling and assaying of lead ores. Also covered is the metallurgical treatment of lead ores, as well as the desilverization of lead by the Pattinson Process and the Parkes Process. Hofman's text has long been considered one of the most important early works on the recovery of silver from lead based ores. 8.5" X 11", 452 ppgs. Retail Price: $29.99

Ore Sampling For Small Scale Miners - First published in 1916, it has been unavailable for over a century and sheds important light on historic methods of ore sampling in hardrock mines. Topics include how to take correct ore samples and the conditions that affect sampling, such as their subdivision and uniformity. Particular detail is given to methods of hand sampling ore bodies by grab sample, pipe sample and coning, as well as sampling by mechanical methods. Also given are insights into the screening, drying and grinding processes to achieve the most consistent sample results and much more. 8.5" X 11", 124 ppgs. Retail Price: $12.99

The Extraction of Silver, Copper and Tin from Ores - First published in 1896, it has been unavailable for over a century and sheds important light on how historic miners recovered silver, copper and tin from their mining operations. The book is split into three sections, including a discussion on the Lixiviation of Silver Ores, the mining and treatment of copper ores as practiced at Tharsis, Spain and the smelting of tin as it was practiced by metallurgists at Pulo Brani, Singapore. Also included is an overview and analysis of these historic metal recovery methods that will be of benefit to those interested in the extraction of silver, copper and tin from small mines. 8.5" X 11", 118 ppgs. Retail Price: $14.99

The Roasting of Gold and Silver Ores - First published in 1880, it has been unavailable for over a century and sheds important light on how historic miners recovered gold and silver rom their mining operations. Topics include details on the most important silver and free milling gold ores, methods of desulphurization of ores, methods of deoxidation, the chlorination of ores, methods and details on roasting gold and silver ores, notes on furnaces and more. Also included are details on numerous methods of gold and silver recovery, including the Ottokar Hofman's Process, the Patera Process, Kiss Process, Augustin Process, Ziervogel Process and others. 8.5" X 11", 178 ppgs. Retail Price: $19.99

The Examination of Mines and Prospects - First published in 1912, it has been unavailable for over a century and sheds important light on how to examine and evaluate hardrock mines, prospects and lode mining claims. Sections include Mining Examinations, Structural Geology, Structural Features of Ore Deposits, Primary Ores and their Distribution, Types of Primary Ore Deposits, Primary Ore Shoots, The Primary Alteration of Wall Rocks, Alterations by Surface Agencies, Residual Ores and their Distribution, Secondary Ores and Ore Shoots and Vein Outcrops. This hard to find information is a must for those who are interested in owning a mine or who already own a lode mining claim and wish to succeed at quartz mining. 8.5" X 11", 250 ppgs. Retail Price: $19.99

Garnets: Their Mining, Milling and Utilization - First published in 1925, it has been unavailable since those days and sheds important light on the mining, milling and utilization of garnets. Included are details on the characteristics of garnets, where they are found and how they were mined. 78 ppgs, 10.99

Gemstones and Precious Stones of North America - Leading mining historian Kerby Jackson introduces us to a classic mining publication on the gems and precious stones of the United States, Canada and mexico. First published in 1890, it has been unavailable since those days and sheds important light on the gems and precious stones that may be found in North America. Included are chapters on diamonds, corundum, sapphire, ruby, topaz, emerald, disapore, spinel, turquoise, tourmaline, garnets, beyrl, peridot, zircon, quartz crystals, feldspars, pearls and many others. Included are details on where these gems and precious stones may be found throughout North America, as well as their characteristics. 360 ppgs, 24.99

Mining Camps and Mining Districts - First released in 1885 by Charles Howard Shinn under the title "Mining Camps: A Study in American Frontier Government", this publication offers a unique look at how early gold miners established their own forms of representative government during the California Gold Rush. Drawing on the the early mining codes of mideviel German miners in the Harz Mountains, on the mining customs of the Cornish tin miners and early Spanish mining laws introduced into California, the miners established the first governments in the American West. 340 ppgs, 24.99

BLM Field Handbook for Mineral Examiners - Leading mining historian Kerby Jackson introduces us to a classic mining publication on mine evaluation. First published in 1962, this work sheds important light on the techniques of BLM Mineral Examiners to perform validity on mining claims. 132 ppgs, 10.99

Six Months In The Gold Mines During The California Gold Rush - Unavailable since 1850, this important work is a first hand account of one "49'ers" personal experience during the great California Gold Rush, shedding important light on one of the most exciting periods in the history of not only California, but also the world. Compiled from journals written between 1847 and 1849 by E. Gould Buffum, a native of New York, "Six Months In The Gold Mines During The California Gold Rush" offers a rare look into the day to day lives of the people who came to California to work in her gold mines when the state was still a great frontier. 8.5" X 11", 290 ppgs. Retail Price: $19.99

The Discovery of Gold in Australia - First published in 1852, it has been unavailable since those days and sheds important light on Australia's gold mining history. Included are rare communications between British agents and the British Crown when gold was first discovered in Australia in 1851. This rare text contains hard to find details on Australia's first mining camps and Britain's early attempts to provide for the orderly regulation of gold mines in that part of the world. Also of interest are hard to find extracts of articles that appeared in the early colonial newspapers that did their best to report on Australia's gold rush as it took place.
102 ppgs, 10.99

www.ingramcontent.com/pod-product-compliance
Lightning Source LLC
Chambersburg PA
CBHW081607200526
45169CB00021B/2211